HOW ANTIGRAVITY BUILT THE PYRAMIDS

The Mysterious Technology of Ancient Superstructures

HOW ANTIGRAVITY BUILT THE PYRAMIDS

NICK REDFERN

This edition first published in 2022 by New Page Books, an imprint of
Red Wheel/Weiser, LLC
With offices at:
65 Parker Street, Suite 7
Newburyport, MA 01950
www.redwheelweiser.com

ISBN: 978-1-63748-002-1
Library of Congress Cataloging-in-Publication Data available upon request.

Cover design by Kathryn Sky-Peck
Cover © Tatyana Maximova | Dreamstime.com
Interior by Steve Amarillo / Urban Design LLC
Typeset in Adobe Garamond Pro and Trajan

Printed in the United States of America
IBI
10 9 8 7 6 5 4 3 2 1

CONTENTS

HOW ANTIGRAVITY BUILT THE PYRAMIDS

INTRODUCTION

Imagine, if you will, the following scenario before you: It's several millennia ago, and deep in the heart of what is now called the Middle East. There are desert sands in every single direction. The temperature is soaring, and it's only getting hotter. And the air is as dry as it could be. Suddenly, the sky above you begins to grow dark; ominously, to be sure. Huge shadows appear here, there, and everywhere. Those same shadows slowly begin to envelop just about all and everyone in sight. They even blot out the light of the Sun itself. You know only all too well from where those massive shadows are coming. The skies. You have seen this happen time and time again, but on each and every occasion you simply cannot fail to be completely awed—hypnotized almost, perhaps—by the incredible view that is above your head. You look up and you squint; it's all you can do.

In all directions, countless huge stones, many of them in excess of 1,000 tons in weight, are slowly moving along above the landscape. That's right: *above.* Like an unstoppable army of determined, giant ants devoted to their queen, the armada of stones follow the paths of those who are controlling them. You continue to stare as the stones finally reach their location. Then, bit by bit, and with shockingly incredible preciseness, the stones are lowered one by one, all into place. Each massive block has been carefully molded into perfect shape and is now becoming part of a temple of absolute gigantic proportions. The process takes only a few days before it is repeated again and again. Neither you, nor any other of your people, have any real, sure ideas as to who the movers of the stone are. And, all of the time, and while the work is going ahead, deep, humming sound penetrates your body, causing it to slightly vibrate.

You have pondered long on who these people, who come from the skies, are: a mighty race from a land faraway on Earth? Gods? Beings from another world in the sky? All that you really know is that one day they *weren't* there. Then, the next day, they *were* there—and in massive, near-worldwide proportions. And their arrival, along with their increasing and growing presence, is changing your world: dotting the planet with huge, stone enigmas that still have yet to be beaten when it comes to the mystery stakes. Now they're even teaching you how to create such structures for you and those of your villages, towns, and cities, such is the ease of how the operations work. Indeed, it turns out that the entire process is amazingly simple to put into place and far quicker than you thought such a thing could be achieved. Eventually, though, the time arrives when those who have all of the power to raise the mighty stones and create massive buildings with amazing, awesome ease, leave. There were rumors and tales of this incredible, mysterious civilization having sculpted unimaginable landscapes in *other* lands and nations. Overseas, even, too. You never forgot them. How could you?

Welcome to the world of what I call the Levitators.

Today, all that is left of this enigmatic civilization—and of their sensational aerial skills—are myths, legends, and untold numbers of massive blocks of stone that forever perplex and tantalize us. How could such incredible technology have been created and put into action with such ease and speed? These questions were asked then. They are still asked now. Although we don't have the full story, we *do* know that the key to moving the immense blocks of stones was nothing less than sound. Yes, really. The Levitators mastered the technology of sound thousands of years ago. What they called the technology back then, we have no idea. Today, however, it's known as acoustic levitation. And for all of our science and our technology, when it comes to it, we're still very much in our infancy, still trying to solve something that we simply just cannot grasp. Not to a significant degree, at least.

Later, we'll get to the deeper side of this amazing science that—for centuries upon centuries—allowed so many incredible, famous structures to be put into place, and so long ago. For now, though, here's a perfect, concise description of acoustic levitation from Marie D. Jones and Larry Flaxman, who have, themselves, sought to solve the mystery of ancient antigravity.

The Jones-Flaxman team present the phenomenon as "two opposing sound frequencies with interfering sound waves, thus creating a resonant zone that allows the levitation to occur. Theoretically, to move a levitating object, simply change or alter the two sound waves and tweak accordingly."[1] You may think that is all very simple. *Too* simple, even. In some ways, it actually is. At least on a very small scale. That's the big problem for us today. The technology allows us to successfully utilize it only on a tiny level. Somehow, though, the ancients were able to use acoustic levitation on an almost unbelievable size.

For us, it's very much like not seeing the forest for the trees.

How Stuff Works says the following, demonstrating that the science is now growing and adding to the science of all this:

> *To understand how acoustic levitation works, you first need to know a little about gravity, air and sound. First, gravity is a force that causes objects to attract one another. The simplest way to understand gravity is through Isaac Newton's law of universal gravitation. This law states that every particle in the universe attracts every other particle. The more massive an object is, the more strongly it attracts other objects. The closer objects are, the more strongly they attract each other. An enormous object, like the Earth, easily attracts objects that are close to it, like apples hanging from trees. Scientists haven't decided exactly what causes this attraction, but they believe it exists everywhere in the universe.*[2]

Could the construction of the world's biggest buildings really have been achieved so incredibly easy? And when we can't even barely get off the starting blocks? For the Levitators, at least, the answer is a decisive *yes,* to both questions. As for us, to a degree we're still fumbling around in the dark. Or, perhaps, *fumbling in the sand* would be a far more appropriate phrase, taking into consideration many of the amazing sites that we will be visiting as this amazing story grows and grows.

For the most part, beliefs in the existence of these vanished, mighty, ancient people—who cracked the secrets of nothing less than antigravity —are largely scoffed at, ignored and dismissed by many, even in some aspects of the scientific community. We're talking about millennia-old, ancient, real-world people similar to those described in the tales of mythical Atlantis and Mu. Or, conversely, we might be talking about honest-to-goodness extraterrestrials. As we continue with our story, the origins of these stonemasons of the absolutely incredible kind, begin to clear.

Did these people—whatever their origins—finally leave our world behind, taking all of their incredible secrets with them? Were those same secrets destroyed, leaving us with only fragments of a science that was radically different to that of ours? Have they periodically come back to share their secrets with a few of us? Or, is everything lost to us? Do we only have, now, myths and legends to work on? All of these questions will be addressed in the pages of this book.

To try to get the answers to this enigmatic riddle, we need to travel much of the globe. That's exactly what we will be doing. We will cross paths with the likes of Stonehenge, the massive stones of Baalbek in Lebanon, the Egyptian pyramids, Easter Island, and numerous other creations, that may very well have been erected in some of the most alternative ways that one could possibly ever imagine. Then, having done that, we'll take a handful of trips through the centuries and then, after all of that, a final great leap to see how we, today, in the 21st century, are now slowly starting to realize just how and why our civilization, and our historians, have gotten so much massively wrong.

CHAPTER 1

WHEN STONES FLOATED HIGH OVER EGYPT

There is no doubt that when it comes to the controversial issue of antigravity—and the ways and means by which it may have been spectacularly achieved thousands of years ago—most researchers of this enigmatic phenomenon tend to head in the direction of Egypt for the answers to the riddle. That, or in the direction of that bastion of data, the internet. You only have to look at the Pyramids to see why the Pyramids of Giza still amaze people after such a long period of time. Largely, it's because the mighty structures are so immense and so unmissable. On top of that, they're so incredibly precise in their construction. And that's why we're beginning with Egypt and its connections to antigravity: it's the ultimate story to get the whole story going.

It's particularly interesting to note that at least some "ancient astronauts" devotees suspected, *decades ago*, that acoustics—of some particular kind—played far more than a few key roles in the maneuvering of mammoth stone blocks high into the skies and into place. One of those who spent a great deal of time thinking about this angle of construction with the stones at Egypt was Peter Kolosimo. No longer with us now, he was the author of such books as *Not of This World* and *Timeless Earth*, and was also someone fascinated by the mysteries of the past. Indeed, he wrote a sizeable number of books. Having studied and translated ancient documentation, Kolosimo said something highly relevant to this whole story—something that is part and parcel to the incredible mystery before us: "According to an Arab legend, the Egyptians used scrolls of papyrus with magic words written on them, on which blocks for the pyramids came flying through the air."[1]

The Pyramids of Egypt: Constructed by Antigravity (Wikimedia Commons)

To have seen such a thing happen before ones' eyes must have been incredible. The word *awestruck* inevitably comes to mind. I need to emphasize, however, to one and all that Kolosimo most definitely did *not* believe in the existence of *literal* papyrus being utilized to help raise massive stones into the sky. He did, however, strongly suspect that the old tales had a fair degree of truth attached to them, but that, over so many centuries, they had been altered—not deliberately, though—via the likes of folktales, Chinese whispers, and legends and myths. Kolosimo was not the only one, however, who went on a limb and addressed this matter of the levitation of Egyptian pyramid stones in distant times. It's time to look at yet another pioneer in this strange world.

There Is Only One Answer to the Riddle

A citizen of New Zealand, Bruce Cathie was both a pilot and a highly driven UFO sleuth, someone who got involved in the UFO field when the flying saucer phenomenon was at its height. Cathie's fascination for the antigravity issue and the UFO topic specifically began in 1952. That was the year in which Cathie had an encounter with a definitive shining, gleaming flying saucer in the skies over Auckland, New Zealand. His life was not just touched by that sighting of the incredible type. It was forever changed. Unlike a lot of ufologists of that era, who were very much focused on what were then-modern-day UFO encounters, Cathie was very much driven by the many puzzles of the past. That's where he thought the answers could be found. Whether he knew it at the beginning of his work or not, the past was indeed where he found himself.

Over time, Cathie came to believe that there was something strange, yet also utterly real, to the rumors of levitation and of massive stones that were said to have been carefully put to the skies in the past. As time advanced and as Cathie, bit by bit, began to refine his theories concerning how UFOs are able to fly in our skies, he came to suspect that our planet was swamped with what amounted to a huge, world-wide "grid." Technology that allowed the mysterious pilots of the UFOs the ability to "ride our skies,"[2] as he put it in his own words.

"Struck by a Rod"

There's another, and far more important, issue, too. It has its ties to Egypt. No surprises there. Just like Peter Kolosimo, Bruce Cathie concluded that levitation, of some kind, was also responsible for the amazing movements of huge stones at Egypt. Clearly, Cathie had been inspired by Kolosimo.

Cathie wrote: "It is inconceivable that the thousands of stone blocks, each weighing many tons, that were used in the construction of the Great Pyramid were dragged hundreds of miles by slaves, then fitted together with such precision that a visiting card cannot be pushed between them. . . . These massive blocks were moved by the utilization of the power grid. . . ."[3]

Kolosimo and Cathie, then, were very much in tandem with each other. That much is made certain when you read the following words of Cathie: "The Arabs have an interesting legend that when the Pyramid was built, the great stones were brought long distances from the quarries. They were laid on pieces of papyrus inscribed with suitable symbols."[4] But, that was not all. There was a very strange aspect to all of this, as Cathie explained. He said of the huge, Egyptian stones that they were "struck by a rod, whereupon they would move through the air the distance of one bow shot . . . there is only one answer to the riddle of such construction methods: Antigravity. Only by this means could the large blocks be moved over great distances and placed so accurately."[5]

You will note that even between Kolosimo and Cathie there were certain, clear differences in the translations concerning (a) Egypt and (b) Antigravity. It should be stressed, too, that those translations are still very much a debate to this very day. That's one of the reasons why we need to be very careful how we interpret the available material that is in our hands.

The Key of Solomon, Witchcraft, and Levitation

There is another aspect to all this. It's tied directly with the legendary *Key of Solomon*. A succinct explanation is required of this legendary tome:

> The Key of Solomon the King: Clavicula Salomonis *is the classic grimoire, or book of magic, incorrectly attributed to the ancient King Solomon who reigned over Israel during the 9th century BC.*

The work likely dates back to the 14th- or 15th-century Italian Renaissance and is a typical work of Renaissance magic. Detailed within this volume are many examples of ritualistic magic typical to the ancient world.[6]

In his 1961 book, *A Treasury of Witchcraft*, Harry E. Wedeck has a section on the phenomenon of antigravity. Under the heading of "Levitation: Magic Flight," there's a very interesting quote that reads as follows from the *Key of Solomon*: "Take the oak rod, turn in the direction in which you want to fly, and write the name of your destination on the ground."[7] That sounds very similar to those earlier legends of rods and papyrus, does it not?

Wedeck also highlighted a fascinating story from early times: "In the third century B.C. Prince Mahendra, Buddhist, accompanied by disciples, went on a mission to Ceylon. He accomplished the journey by levitation. Rising into the air and alighting in Ceylon on the Missa Mountain, where Buddhist temples were later established."[8] Wedeck made it clear in his classic book that he was sure that levitation was not just a myth. It was the genuine, priceless item.

CHAPTER 2

MIND-POWER AND EFFORTLESS CONSTRUCTION

Now our story jumps from the pyramids to the Pentagon. It's a brief jump, though. And, yes, I do mean *that* Pentagon: nothing less than the HQ of the United States Department of Defense (DOD). In a very strange fashion, the two blended together, despite the differences of time. I need to stress that what you're about to read now is without doubt somewhat of a diversion—and, as well as that, it sets the scene for something that will, as the story grows and grows, bring us right back to Egypt and levitation.

Here's my story, how it began, and finally how it came to its peak. In 2006, I was looking for a new subject to write about. The challenge for me was: what should it be? Well, that was the million-dollar question. As fate would have it, however, an amazing story—one that tied in with much of this book—fell right into my lap. It came from a UFO investigator who was also a priest. His name is Ray Boeche. He's a good friend, too.

The Story Really Takes Off

To give you an idea of his work, I'll share with you parts of his biography:

> *Founder and former director of the Fortean Research Center, The Reverend Dr. Raymond W. Boeche has been involved in the study of unexplained phenomena since 1965. He has served as Nebraska State Director for the Mutual UFO Network, on the Board of Advisors for Citizens Against UFO Secrecy, and in various capacities with numerous other organizations around the world, involved in the study of unexplained phenomena. . . . A graphic artist and book designer with the prestigious University of Nebraska Press for nearly twenty years before retiring, Dr. Boeche holds a B.A. from Peru State College, a Th.M. degree from St. Mark's School of Divinity, and a ThD. from St. Paul Theological College.*[1]

As Boeche and I were chatting, I brought up the issue of what I might write about for my next book. To be honest, the project I originally planned on doing wasn't that great. *C'est la vie.* Boeche, however, said that he had an incredible story to tell, something I would doubtless find fascinating. Sounded good! Boeche explained to me that the story had been in somewhat of a limbo for more than a few years, but he quickly filled me in. And I'm very pleased that he did. Boeche did more than just share the story with me: he also suggested that I should try and crack the mystery open. *Wide open.*

It all circulated around a secret, small group that was buried deep within the Pentagon. And, that same group was quietly investigating something that was both amazing and disturbing. It was the theory that the UFO phenomenon is not extraterrestrial, but outright demonic in nature: Heaven, Hell, demons, the fiery pit—the whole thing. And, because of his deep connections to both religion *and* UFOs, Boeche was rightly seen by the Department of Defense as someone who could help the U.S. government get its grips into this very weird story. When Boeche told me the saga, I knew *this* was the one.

The Story Was Unleashed—and Maybe the Demons Were, Too

On the morning of January 22, 2007, I interviewed Ray Boeche about his research,[2] that mysterious Pentagon group, and the theory that aliens were far from what they claimed to be: the deadly and dangerous minions of the Devil himself. For Boeche, it all began in November 1991. This part of the story demonstrates just how long it had been quietly circulating in the realm of Ufology, but—somewhat puzzling—with very little having been done with it. In fact, almost none.

November 25, 1991 was the day on which a pair of U.S. Department of Defense physicists met Boeche for lunch, and for a chat that lasted for more than three hours, at what is now titled the Lincoln Marriott Cornhusker Hotel, Nebraska. The pair was deeply concerned—worried and, maybe, even frightened—as to what was going down in the heart of the government. And just how far things might go—perhaps, to the point of no return. At the time, the two men were working on a DOD project to try and directly communicate with dangerous, supernatural creatures. The DOD gave them the somewhat bland title of *nonhuman entities* (NHEs). Granted, that's exactly what they were, but there was much more to them, as you'll see as our story grows and grows. For some employees of the U.S. government, the NHEs were aliens. For others? Full-on demons. More than a few didn't know what to think—only that the project was determined to find the truth of the UFO enigma.

Photos of the Grim Kind and Deadly Activity

As Ray Boeche ate his free lunch and listened carefully, a grim story proceeded to get even grimmer. People who were working on the project tried to make "deals" with the NHEs. Yes, really: that's just how crazy and dangerous things got. Unfortunately, the demons, if that's what they were, were far more powerful and manipulative than us. People in the

DOD project began to fall sick, both physically and mentally. Near-endless bad luck blighted nearly all of the group. It was as if a mass of dark clouds had descended upon one and all who had dared to get involved. Some of the team quietly thought they had all been cursed by the Devil himself. Perhaps that's what really happened. Wildest of all, the DOD operation was trying its best to find ways to weaponize the powers of the nonhuman entities. Dealing with the Devil to make use of his very own arsenal of supernatural weaponry? It sure sounded just like that, as bizarre as it seemed. Or, as Boeche worded it, "a vast spiritual deception."

Then, there were the deaths.

Close Encounters of the Fatal Kind

The most disturbing aspect of all of this, Boeche said to me, was that a number of attempts to try and contact the NHEs resulted in a number of sinister and strange deaths. Whether this occurred within the Pentagon, or at another government facility, the two physicists didn't tell Boeche. Or, they *wouldn't* tell. I asked Boeche to share with me this grim part of the story. He did exactly that to the extent that he felt he could. One of the men at the Cornhusker Hotel brought out of his briefcase an envelope that contained large photos of three people. Clearly, they were all dead. Their fresh corpses were sitting in large seats that looked not unlike a dentist's chair.

It was made clear to Boeche that the deaths had occurred while certain experiments were running wild. Those experiments were designed to try to contact the NHEs by using what is popularly known as remote viewing. Indeed, I have to make it clear that the NHEs did not just manifest out of nowhere in the heart of the Pentagon. All of the communication with those cold creatures was done in a mind-to-mind fashion.

All three of the bodies had EEG and EKG leads attached to them. One of the dead looked as if he had a "dent" in the side of his head, Boeche recalled. How such a thing happened, the DOD had no idea. Were they volunteers? Prisoners forced to go along with the project? Boeche was never told that side of the story. He still has no idea, to this very day.

That wasn't the end of the story, however.

The Controversial "Often" Story

The more I looked into the story of the Department of Defense group, and their undeniably reckless actions, the more and more I learned about their reckless agenda. In various ways, I found out that the group had actually been around for years; it wasn't a new program, as I had initially thought, and as Boeche had thought, too. The truth was that it was a relatively small program and they had given it the nickname of the Collins Elite. I should make it clear that I never learned of the real, classified name of the project. Neither did Boeche. There was such a secret name, though; that much was told to me. There was something else, too: the Collins Elite had inherited a number of files from a similar group in the Central Intelligence Agency (CIA). It was titled Operation Often.

It was run by a man named Sidney Gottlieb. He was a chemist who, for years, in the 1960s and 1970s, worked on the CIA's notorious MK-Ultra "mind control" project. It, too, was trying to find ways to weaponize various aspects of the paranormal, the supernatural, and the occult: Ouija boards, Voodoo, curses, magic, witchcraft, and more, even something along the lines of assassination by the use of occult powers. The only difference between the Collins Elite and Operation Often was that the latter operation wasn't working on anything to do with UFOs or aliens.

Among those aforementioned files that were handed over to the Collins Elite years ago were various U.S. government documents that dated back to the 1950s. They told an incredible story. It was a story of acoustic levitation, no less. Now, you will see why I had to make such a diversion away from Egypt to the heart of America's Department of Defense. And, as it happened, back again.

CHAPTER 3

EGYPT, HALLUCINOGENS, AND THE SECRETS OF THE PAST

The next part of this undeniably strange story very much falls into the category of government secrets, but also into the domains of chemical hallucinogens and mushrooms. Egypt, too. And utterly effortless acoustic levitation. To say all of that is quite a heady mix is not wrong. Now, I'll bring us up to how, in the 1950s, a controversy-filled attempt—that involved some of the early staff of what would become the Collins Elite, no less—was quietly put into place for the Department of Defense. It was to remote view nothing less than Egypt's famous Giza Plateau. And what, exactly, did the psychic team find? I'll tell you: incredible images of Egypt projected into their minds, Egypt from thousands of years ago, rather than Egypt of the 1950s.

On top of that, massive stones were seen by the remote-viewing team moving through the skies of the desert land, albeit not at high levels at all, and at slow speeds. Torturously slow speeds at some times, even. Nevertheless, those who were directing the stones—with what looked to be seven-to-eight-feet-long metal poles of some sort—had complete control over the mighty blocks. Just touching the stones with the poles caused them to become incredibly light. We are, of course, talking about acoustic levitation. This is all very much akin to what Peter Kolosimo and Bruce Cathie had got their grips into.

Was it all just a coincidence that Cathie started looking into the matter of Egyptian pyramids and antigravity at the very same time as the remote-viewing (the 1950s)? Or is there a bigger story to this—a story that still has yet to surface? Those questions bring us to the full story of how and why the Department of Defense, back in the Fifties, chose to take things to *another* level and have its most skillful psychics penetrate the very heart of ancient Egypt even further. And it all circles around what is known as *Amanita muscaria.*

From Stones to Stoned

The U.S. Forest Service arm of the United States government says of *Amanita muscaria* that: "In the 'old world,' the psychoactive fly agaric mushroom (*Amanita muscaria*) has been closely associated with northern European and Asiatic shamans and their rituals. Researchers have documented its use or presumed use by numerous cultures throughout Europe and Asia. In Siberia, its use predates the crossing of the Bering Straits into North America."[1]

The government says more on this fascinating topic:

> *During the ceremonial ritual, the shaman would consume and share the sacred mushrooms with the participants. The smoke hole was a gateway or portal into the spiritual world where the people would experience many visions. Among the Sami (Laplander) peoples, the hallucinations associated with ingestion of fly agaric gave the sensation of flying in a 'spiritual sleigh' pulled by reindeer or horses (i.e., Santa in his sleigh journeying out into the night to give gifts).*[2]

Anomalous Cognition or Second Sight

It's well-known that since the 1950s, the U.S. government, the Defense Intelligence Agency, the FBI, and the CIA have done their absolute best to find ways to psychically spy on our enemies. And the Russians did the same, making us their targets. Now, it's largely terrorists. We had one description of remote viewing above, thanks to Simeon Hein. Here's another concise description of it, a description that adds more to the nature of the strange phenomenon. From *Gaia* we have this:

> *Remote viewing is defined as the ability to acquire accurate information about a distant or non-local place, person or event without using your physical senses or any other obvious means. It's associated with the idea of clairvoyance, seemingly being able to spontaneously know something without actually knowing how you got the information. It is also sometimes called "anomalous cognition" or "second sight."*[3]

In the 1950s, certain elements of the U.S. government quietly chose to experiment even further with this particularly potent hallucinogenic. That's right: we're talking about a top-secret program that came on the scene long before the work of the CIA remote viewers of the 1970s were anywhere in sight.

A Man Called Puharich

The next part of our story—of pyramids, of the Collins Elite, of Egypt, and of much more—is largely aimed in the direction of two, specific people: Andrija Puharich and Harry Stone. Puharich was American and had a Yugoslavian background. His lab—*also* in the 1950s, interestingly—soon became an absolute hotbed for psychic work. The location was Glen Cove, Maine. Puharich also had another card in his deck. At the time in question, he was nothing less than a captain in the U.S. Army, someone who was firmly attached to the military's Army Chemical Center, located in Maryland. For two years, from 1953 to a portion of 1955, Puharich did his very best to try to find how to use the human mind to spy on the Russians, and to do so by using psychic means.

To some degree, at least, there was a crosspollination between the work that Puharich was up to and that which the CIA's mind-control program known as MK-Ultra was secretly running. It was inevitable, then, that the two would come together. They did.

Mushrooms of the Sacred Type

It's a fact that Puharich had a fascination for all-things of an Egyptian nature. And, as a result of that, on one occasion in 1955 (that eventually became *many* occasions), Puharich chose one particular man to use his mind—mixed with a fair amount of *Amanita muscaria*—to try to take a view of Egypt as it once was all of those millennia ago. Way back when, is what we're talking about. That man who Puharich brought into the fold was one Harry Stone. He was a man whose psychic work revealed amazing data on Egypt and levitation. Stone would also, a few years later, become one of the key players in Puharich's 1959 book, *The Sacred Mushroom: Key to the Door of Eternity*. The blurb from Puharich's book reads as follows:

> *The book describes in vivid detail Puharich's extraordinary association with Harry Stone, a young sculptor of unusually acute extrasensory perception, who on a number of occasions spontaneously went into a deep trance state and then began to speak and write in the ancient Egyptian language. Identifying himself as Ra Ho Tep, a high-born Egyptian who lived 4,600 years ago, he defined the long lost ritual of the sacred mushroom and its astonishing effects upon the human consciousness.*[4]

The Visions Get Even More Amazing

There was more to come. In his various hallucinogenic states, Stone saw, thousands of years earlier, those massive stones floating across the skies at heights of anywhere from roughly twenty feet to around five hundred feet high, depending on where and when the stones were needed to be placed. All of the stones were designed to be carefully placed next to each other—albeit without a single bit of human muscle-power required. Iron rods, of a very strange and mysterious type, were all that were needed to get the incredible jobs done. On every occasion, Stone came out of his confused and befuddled mindset—something that is not surprising.

It's quite reasonable to say that while under the effects of *Amanita muscaria* Harry Stone could have seen just about *anything*. Moreover, and after all, he did admittedly have a *pre-existing* fascination for the history and the legends of Egypt. Stone even described himself, in an earlier incarnation, as a high-born Egyptian. That he should have specifically seen incredible images of massive, floating, multi-ton stones traveling over Egypt, however, strongly suggests that Stone was on the very same wavelength that we are now—and that Bruce Cathie and Peter Kolosimo would also come to.

Remote Viewing and the Planet Mars Connection

So what we have here is a strange, and somewhat complicated, story of the demon-obsessed Collins Elite and demons, remote viewing, the CIA, Operation Often, and what was *really* going down in ancient Egypt, all those years ago. And all ending up in one big, amazing mix.

Thus, we have the story—and, some might dare to say, even the facts—behind the secrets of the construction of the Pyramids by mysterious iron poles. We are not quite finished with Egypt, though. There's another incredible structure that needs to be put into the spotlight: the Sphinx. It, too, is swamped in mystery and altered history. We'll get to it in the next chapter.

Finally, though, for this particular chapter, you might think that the story is just too far out to be true. Wrong. Consider this: thanks to the provisions of the U.S. Freedom of Information Act, we know that something fascinating—and deeply relevant to this overall story—happened on May 22, 1984. That was the date on which the Central Intelligence Agency secretly remote viewed the planet Mars. We don't really know why, which makes the CIA story even more thought-provoking.

Other viewings occurred in early 1985. On several occasions, the team aimed their targets at what looked like huge pyramids on Mars. That both Harry Stone (in the 1950s) and the CIA's remote viewers (in the 1980s) were ordered to try to figure out the truth of the Pyramids—whether of Mars, or of our planet—suggests someone (or some agency) knows far more about our ancient history than we do.

Now, it's Sphinx time.

CHAPTER 4

THE SPHINX: NOT AT ALL WHAT IT SEEMS TO BE

The story that you are about to read tells of one of the most iconic, well-known creations in known human civilization: the Sphinx of Egypt. Convention tells us that the Sphinx was created somewhere roughly around 2575 BC and 2465 BC. Convention, however, may be completely wrong. There are incredibly important reasons why that could be the case. In fact, there is a distinct possibility that the Sphinx was created as far back as *12,000 years ago*[1]—as you will soon see. Yes, you did read that right. And, no, I didn't hit a couple of wrong "zero" keys while I was writing this book. Once again, history is about to be altered.

What we know for sure about the Sphinx are its dimensions. It's an impressive almost-240-feet-long, 66-feet-tall, and 62-feet-wide creation. See it once in person and you'll never forget it. The problem that we have—and, undeniably, it's a big problem—is that if the Sphinx *was* indeed built around 12,000 years ago, then this radically changes much of the conventional scenario of Egypt's history. In fact, such a thing would alter history forever. Yet, there is a high degree of merit to this controversy. At the very least, it suggests that there were highly developed Levitators in Egypt long before we have all been told. There's nothing wrong about that, at all. It just means that significant numbers of people, in the world of Egyptology, have got their timelines spectacularly wrong.

Writing for *Live Science*, Owen Jarus provides us this:

"Villages dependent on agriculture began to appear in Egypt about 7,000 years ago, and the civilization's earliest written inscriptions date back about 5,200 years; they discuss the early rulers of Egypt."[2]

The word *unclear* is a most relevant one to apply to this particular story.

The *Real* Creators of the Sphinx Were, and Still Are, Completely Unknown to Us

Note that even the "7,000 years ago" that Owen Jarus presents to us is still nowhere near 12,000 years ago. That still leaves us with a time frame of 5,000 years to address and solve. There are other issues to be addressed, too, all of them controversial and, yet, equally amazing. Historians and archaeologists will tell you that the Sphinx was the work of the people who lived when one Khafre ruled. You might very well ask: who was he? Khafre was the pharaoh who ruled in what is known in Egypt as the Old Kingdom. The staff of the Metropolitan Museum of Art (New York) put the date for the Old Kingdom at somewhere around 2649 BC to 2130 BC. Khafre was the fourth king of the fourth dynasty.[3] He is acknowledged as having overseen the building of the second of the three Giza pyramids. There is, however, an altogether fascinating, but completely different, scenario for me to share with you. A highly alternative scenario suggests Khafre, in essence, inherited the Sphinx, rather than having had it constructed to his personal orders. And, in doing so, Khafre had the incredibly ancient Sphinx reworked to his very own demands.

This provokes an amazing, and even beyond incredible, picture: that Khafre, himself, may never have known who the mysterious creators of the giant Sphinx really were. That suggests that the *real* creators of the Sphinx were, and still are, completely unknown to us. And that they were around thousands upon thousands of years before Khafre and his people were ever on the scene. Such a scenario is actually not at all impossible, despite what some might say to the opposite. After all, think about it: Khafre really only had to have the head of the Sphinx carefully changed to ensure it fitted his very own appearance. As a result, Khafre had a near-ready-made construction of all his very own. We, up until very recent times, knew nothing about it. There was one man who suspected that the Sphinx was older than most assumed, someone who truly went against the grain and became a hero in the world of what we might term "alternative history."

The Sphinx: A definitive enigma (Wikimedia Commons)

Before we get to that man, there's *another* one who *also* suspected that the Sphinx was much older than we are told. Let's give him the kudos he deserved.

Many Excavations throughout Egypt

Although this particular theory concerning the Sphinx's origins was not a particularly popular scenario, it has certainly had its supporters over the decades. One of them was an archaeologist by the name of Selim Hassan. In fact, as far back as the 1940s he was already touting that particular, controversial picture of a *really* ancient Sphinx.

So what we have here is a highly regarded figure in the world of Egyptology, someone who suggested that our conclusions concerning the Sphinx were way off the mark. Yet so many people in this particular field turned their heads completely the other way. They shouldn't have. It is, after all, their loss.

Now, we have to come to the one person—more than anyone else—who has made an incredible, extremely strong argument that the Sphinx most likely had nothing to do with Khafre and/or the Old Kingdom. That man is Robert M. Schoch, PhD. By the time you've read this chapter, you might consider him as a maverick. It is, however, the mavericks of this world who get things done. Schoch definitely did that.

The Strange Phenomenon of Egypt's Incredible Rainfall in Times Far behind Us

Robert Schoch has a most impressive background. He is

> *a full-time faculty member at the College of General Studies at Boston University and a recipient of its Peyton Richter Award for interdisciplinary teaching. Schoch earned his Ph.D. in Geology and Geophysics at Yale University in 1983. He also holds an M.S. and M.Phil. in Geology and Geophysics from Yale, as well as degrees in Anthropology (B.A.) and Geology (B.S.) from George Washington University.*[4]

That's to say, Schoch most assuredly knows his history. As you'll see now, he also knows his alternative takes on history. That includes the theory suggesting that the Sphinx was constructed about 12,000 years ago. It's intriguing to note that much of the theorizing on the part of Schoch comes from the matter of nothing less than rain—specifically, Egypt's rainfall and water erosion on the Sphinx. That may sound totally strange, yet Schoch makes a great, solid case for his argument that spectacular errors have been made when it comes to the history of the Sphinx. Also, Schoch has come to the amazing, but highly plausible, conclusion that at some point in its history, Egypt had huge amounts of rainfall. Since we know that the rainfall levels of the period of Khafre weren't particularly huge, then a solid scenario can be made that the rain that caused significant water erosion on the Sphinx, which is most definitely in sight if one looks carefully, had to have happened long before those in the field of Egyptology have told us.

12,000 Years Ago: The True Age of the Sphinx?

In his very own fascinating way, Schoch has made a totally plausible case that the Sphinx was indeed the creation of a mysterious, and now lost, people. Highly advanced ones, too. To create the Sphinx 12,000 years ago would have required a huge effort on the part of the ancient people who we know, for sure, existed at that time.

I should stress that Schoch has not dug too deeply into the matter of who, specifically, may have been the original creators of the Sphinx all those millennia ago. Yet, he does make it clear that there had to have been advanced people around at that ancient time, if we are going to go with this incredible scenario. These mysterious people, not Khafre, were the real creators of the legendary, world-famous Sphinx. By now, you've probably come to the same conclusion as me: that the real makers of the Sphinx were none other than the Levitators. Khafre was nothing but a Johnny-come-lately.

We now get to one of the most significant parts of this particular story. It brings us back again to the era of Khafre, the man whose place in history is in drastic need of significant changing. We have to get back to that matter of water. Lots of it. It's most important to note that when Khafre was in power in Egypt, the rainfall levels were hardly what one could term deluge-like. In fact, the absolute opposite was the case. Egypt's rainfall, back then, was really not much different from what Egypt's rainfall is now. You don't see today's Egypt overwhelmed by rain.[5] So, this provokes a very significant question for all of us: how can it be that the Sphinx appears to have suffered from significant amounts of rain? And a very long time ago? That's the question Schoch chose to tackle. In doing so he found the answer.

Schoch himself says of all this: "Seismic data demonstrating the depth of weathering below the floor of the Sphinx Enclosure, based on my analyses (calibrated very conservatively), gives a minimum age of at least 7,000 years ago for the core body of the Sphinx (and more realistically, *on the order of 12,000 years ago* [italics mine])."[6] Schoch goes further: "Standing water in the Sphinx Enclosure would not accelerate the depth of weathering below the floor of the enclosure."[7]

How the Sphinx Looked *Then*, and How it Appears *Now*

It must be admitted that Schoch's theory has not been wholly embraced by mainstream figures in the field of Egyptology. It's not surprising: cherished theories, in the world of Egyptology, are hard to deny or destroy. The same goes for petty egos in that same particular field. That goes for history and archaeology, too. But, I say: So what? The conventional scenario of Khafre being the creator still remains a problem, thanks to Schoch. And it's a big problem, too, when we now know that Egypt was, apparently, once a place with significant amounts of water. There's no doubt that the weathering of the Sphinx *was* caused by incredible amounts of rain and for a huge amount of time.

In light of Schoch's findings, it's very difficult to present any other kind of scenario. The only plausible answer to this riddle is that the Sphinx is so incredibly weathered because, despite what the naysayers stubbornly claim over and over, it was created in an era long before Khafre was ever around—in fact, thousands of years before that ego-driven pharaoh was anywhere around.

Schoch's careful studies of the Sphinx suggest that its head—when it was in its original form, that is—was that of a lion, rather than that of a man. As well as that, it was Khafre's people who did the careful resculpting of the stone head from animal to man. Schoch uses words including *shrunken* and *mutilated* to describe how the Sphinx looked *then*, and how it appears *now.*[8] It says a great deal about Khafre's self-importance that he felt a need to change such a huge structure to something completely else. We are, admittedly, however, still unsure who this ancient civilization, which was creating massive, incredible structures about 12,000 years ago, really were. I still say the Levitators, though.

We Still Don't Know How the Ancient Egyptians Lifted Blocks Weighing Hundreds of Tons

The reason why there is *still* so much debate about how the Giza pyramid complex—or, rather, the Giza Necropolis—was constructed is very simple: the way the pyramids were put together *still* puzzles so many people. *Heritage Science* states: "Pharaoh Cheops (Khufu) began the first project of the Pyramid of Giza, around 2550 BC. Its largest pyramid is the largest in Giza and is about 481 ft. (147 m) high above the plateau. Its stone masses estimated at approximately 2.3 million, weigh an average of 2.5 to 15 tons."[9]

In light of those amazing dimensions, should we believe that such an incredible, collective task could have been achieved by using nothing but muscles, ropes, slopes, ramps, and pulleys? Santa Claus and his reindeer, maybe? Or, should we focus our attentions on that strange "papyrus" and those curious metal "rods" that seemingly made the whole job as easy as pie? Kara Cooney, a professor of Egyptian art and architecture at the University of California, Los Angeles, says: "We actually don't know [the] mechanism of cutting hard stones like red granite. And we still don't know how the ancient Egyptians lifted blocks weighing hundreds of tons up the sides of the pyramids."[10] Amen, to that.

No doubt, the secrets of Egypt will continue, bit by bit, to reveal yet further secrets. Right now, however, *the* most important development *still* remains the one that was found by Robert Schoch.

CHAPTER 5

AN ISLAND OF ENIGMAS

Just like all of the cases highlighted in the pages of this book, the saga of the huge stones of Easter Island is steeped in overwhelming mystery and intrigue. There are some things we *do* know, and there are other parts of the story that we clearly do *not* have a clear angle on. Not at all. The reason is we don't know the full history of the island or its people. That doesn't, though, stop us from getting deep into the controversy surrounding the stones. Let's get right into the story.

As for the island itself, it's situated in the South Pacific Ocean, more than 2,000 miles from Chile, and about 2,500 miles from Tahiti. As for its size, it amounts to slightly more than 60 square miles. For such a small island it has, quite rightly, become such a historic, amazing place. Created out of a number of volcanoes in the distant past, and with numerous caves that reach well into the island's mountains, and with Mount Terevaka being more than 1,600 feet above sea level, we have a fascinating, yet compact, place.

Britannica says:

> *The island's population represents the easternmost settlement of a basically Polynesian subgroup that probably derived from the Marquesas group. The original Rapa Nui vocabulary has been lost except for some mixed Polynesian and non-Polynesian*

words recorded before the Tahitian dialect was introduced to the decimated population by missionaries in 1864. Today Spanish is generally spoken. In their traditions, the islanders consistently divide themselves into descendants of two distinct ethnic groups, the "Long-Ears" and the "Short-Ears." Intermarriage is common, and an influx of foreign blood has become increasingly dominant in recent years.[1]

Much Speculation about the Exact Purpose of the Statues

Let us take a look at the mysterious history of Easter Island and its people. It's a fascinating story. And it has an acoustic levitation aspect to it. First and foremost, it's important that we pay homage to the history of this enchanting island, of its history, and of its people. For example, we should take note of the real name of the island, the one that the people of the island use: Rapa Nui. Easter Island, also referred to as Paaseiland, was given its famous name by explorers from Holland who arrived in 1722, chiefly overseen by one Admiral Jacob Roggeveen. The time of arrival: Easter Sunday. And, that is why, and how, the island got its now-famous name. Although "Easter Island" is a famous title, it's a shame that a bunch of white men from the other side of the world decided that it was their role to make a significant change to a culture that was doing their own thing, and wanted it kept that way.

What else do we know? A great deal. The editors of History.com say:

It was annexed by Chile in the late 19th century and now maintains an economy based largely on tourism. Easter Island's most dramatic claim to fame is an array of almost 900 giant stone figures that date back many centuries. The statues reveal their creators to be master craftsmen and engineers, and are distinctive among other stone sculptures found in Polynesian cultures. There has been much speculation about the exact purpose of the statues, the role they played in the ancient civilization of Easter Island and the way they may have been constructed and transported.[2]

The wonders of Easter Island (Wikimedia Commons)

It was roughly in the period 700 to 800 AD when the first people arrived on Easter Island, long before Roggeveen and his people appeared and caused chaos. The first ruler of the island was one Hoto-Matua, who many historians and archaeologists estimate came to the island with his people. And an enchanting civilization grew. What caused the Europeans to create so much havoc? We don't know the full reason. But we still know a great deal.

Undiscovered Lands

We know that they made shore at what is called Anakena. The location was perfect (at least for those who practically sought to take the place over and without even asking): Much of the coast of Easter Island is nothing but rock, whereas Anakena has a beachy environment, thus making it much easier for the first groups to make a safe landing. Still, on this particular point, Liesl Clark writes:

> *According to an Easter Island legend, some 1,500 years ago a Polynesian chief named Hotu Matu'a ("The Great Parent") sailed here in a double canoe from an unknown Polynesian island with his wife and extended family. He may have been a great navigator, looking for new lands for his people to inhabit, or he may have been fleeing a land rife with warfare. Early Polynesian settlers had many motivations for seeking new islands across perilous oceans. It's clear that they were willing to risk their lives to find undiscovered lands. Hotu Matu'a and his family landed on Easter Island at Anakena Beach. Te-Pito-te-Henua, "end of the land," or "land's end," is an early name for the island.*[3]

The Easter Island Creations Provoke an Air of Mystery

As for the population of Easter Island, at the time the Dutch team of Admiral Jacob Roggeveen arrived and made ground, the numbers were roughly around 2,000 to 3,000.[4] By the late 1800s, tragically, the numbers were way down: to just slightly more than one hundred—of course, all due to those Europeans bringing disease and illness with them, something that could not be combated by the largely isolated people of the island. Very sad—made even more so, because such a terrible situation could easily have been completely avoided: the people of Easter Island could have been left alone.

To be sure, it was a totally disastrous affair. The whole thing graphically demonstrates what can happen when one culture descends upon another, with little thought (or even *no* thought at all) for the people who are *already* in place—and who had been there for a long, long time. On top of that, Easter Island was converted by Catholics who, arrogantly, presumed that it was their absolute right to turn the religion of the people of Easter Island upside down. It was *not* the right of the Catholics to do what the hell they wished to do. That is, however, precisely what they did. Then, not long before the end of the 19th century, Easter Island was annexed by the government of Chile. To say that Easter Island and its people got a bad deal from the western world is a major understatement of epic proportions. Today, however, the population of the island is pushing toward 8,000;[5] a good, solid, healthy number, thankfully.

The Movers of Easter Island

All of this brings us to why, precisely, Easter Island is so well-known and rightly revered. It is, of course, those many enigmatic stones that can be found across the island. Just like Stonehenge, Avebury, Egypt, and elsewhere, the Easter Island creations provoke an air of mystery, and of incredible times and people—and histories—long gone. Also, of ancient secrets and the mysteries that still surround the creation and moving of the stones.

As for those incredible stones themselves, which the people of the island call the "Moai," there are approximately 900 of them. All are skillfully carved and facing out to the all-expansive ocean. Interestingly, when it comes to their size and weight, most of the Moai are around 13 feet high and roughly 14 tons in weight. As for the issue of why so much effort went into all of this, well, the answers to those questions very much elude us. If the island hadn't been plunged into chaos when the Dutch arrived, we might have far more answers than we do now. Sadly, we can't go back and change things. It's something that is overwhelmingly frustrating for those who have spent their time studying the stones. I said that *most* of the stones were of similar weights for a reason. That's correct. One particular Moai, however, really stands out from all of the other. It stands high, too—*way* high. It's highly appropriate title? El Gigante. It couldn't have been better.

The Amazing Moai that Walked from the Quarry

This particular Moai, *Atlas Obscura* reveals, is

> *located near the Rano Raraku Quarry, which stands some 72 feet tall . . . El Gigante weighs in at an astonishing 160-182 metric tons, more than the weight of two full [Boeing] 737 airplanes. However, El Gigante was ambitious even for the master movers of Easter Island. Experts believe that had they finished this Moai (there is some question as to whether they ever intended to), it is unlikely the islanders would have been able to move it. In comparison, Paro, the largest Moai ever erected, is 10 meters (33 ft) high, and weighs 75 metric tons.*[6]

It's no surprise at all that the incredible weights and the huge sizes of the Moai—particularly so when it comes to El Gigante—still baffle people, as to how the massive stones were carefully put into place and, perhaps even more importantly, *why* they were put into place.

On this particular issue, *Thought Co.* says these words:

> *Since the 1950s, various experiments moving Moai replicas have been attempted by methods like using wooden sleds to drag them around. Some scholars argued that using palm trees for this process deforested the island; however, that theory has been debunked for many reasons. The most recent and successful Moai moving experiment, in 2013, involved a team of archaeologists wielding ropes to rock a replica statue down the road as it stood erect. Such a method echoes what the oral traditions on Rapa Nui tell us; local legends say.*[7]

Mysteries of the stones (Wikimedia Commons)

Were the Levitators at work at Easter Island? In my view, yes, they most assuredly were. It was all down to the things that the Levitators did so very well: sculpt an incredibly large number of massive stones, and then carefully place them on some of the most awkward locations available to them. Let's learn more about this fascinating story.

Stones of Incredible Weight

The Khan Academy states: "Around 1500 C.E. the practice of constructing *Moai* peaked, and from around 1600 C.E. statues began to be toppled, sporadically. The island's fragile ecosystem had been pushed beyond what was sustainable. Over time only sea birds remained, nesting on safer offshore rocks and islands. As these changes occurred, so too did the Rapanui religion alter—to the birdman religion."[8]

The mysteries, the secrets, and the reasons why the people of Easter Island were undeniably driven to create so many stone structures—on many occasions, and sometimes in extremely awkward areas of Easter

Island's landscape—remain unclear. All we can say for sure is that this story demonstrates that *yet another* culture, using massive stones, and of incredible weight, and centuries ago, achieved something incredible, something that still perplexes us.

An Unbelievable Achievement

Let's see what other investigators, now gone from this world, thought of the phenomenon of the massive, mysterious stones that can be found on Easter Island. English archaeologist and anthropologist Kathryn Routledge, who, along with her husband, William, made an expedition to Easter Island, arriving in March 1914, said: "In Easter Island the past is the present. It is impossible to escape from it; the inhabitants of today are less real than the men who have gone. The shadows of the departed builders still possess the land. Voluntarily or involuntarily, the sojourner must hold commune with these old workers. The whole air vibrates with a vast purpose and energy that is no more. What was it? Why was it?"[9] Questions upon questions.

Charles E. Sellier, an author and the creator of a number of television miniseries and documentaries, including 1976's *In Search of Noah's Ark,* said something highly valid about Easter Island:

> *Many of these Moai were carefully fitted with red granite "topknots" or head coverings that are separately carved from an entirely different kind of rock and appear to have been put in place after the statue was erected—an unbelievable achievement when you consider these topknots weight from five to eleven tons by themselves. The fact is that what the ancient Easter Islanders did would be difficult to accomplish even with modern power equipment.*[10]

Indeed. Yet the people of Easter Island—or the Levitators—achieved their goals very easily.

Getting Close to Signing Off with Wise Words

Dr. Evan Llewellyn, a historical sociologist, has said:

> *Every culture since time immemorial has raised statues to their gods and personages. The organized social structure of their society was religious, with a well-defined upper class that called themselves Hanau Eepe, or "Long Ears," because they extended their ear lobes with heavy ornamentation. The lower classes were called Hanau Momoko, or "Short Ears." It is revealing that all of the stone heads are of the "Long Ears."*[11]

Elizabeth R. Wheaton, a science librarian, says: "It is quite true that the giant statues depict the 'Long Ears' only, but recent archaeological discoveries have revealed skeletons of a much larger-boned culture, which would seem to support the legend of an earlier race of people on Easter Island."[12]

Sellier's last words on this issue are: "A complete understanding of the history and people of Easter Island seems to be beyond our reach. Perhaps it is precisely that realization that keeps us looking for answers and makes the many mysteries of Easter Island worth investigating."[13]

It's hard to disagree with those words. Still on the matter of words, the following are well worth noting because they make it abundantly clear that the mystery surrounding so many ancient structures comes across as near-magical, when it comes to ancient times. Particularly so at Easter Island.

The Story Continues

There's one particularly fascinating, last issue about the history of Easter Island, something that is also sad and unfortunate. The *Easter Island Travel* company make that abundantly obvious with these particular words:

Rongo-rongo (roŋo-roŋo in Rapa Nui) is an ancient Easter Island glyph writing. It is the only known native writing in all of Polynesia. Rongo-rongo uses symbols of items, as with the Egyptian hieroglyphs. The rongo-rongo symbols were written on tablets of wood. Today, only around 25 rongo-rongo tablets are known to exist; all scattered at museums outside of Easter Island. In 1862-1863, many slave raiders attacked Rapa Nui. All able-bodied men were taken, among them all the wise men who knew how to read and write rongo-rongo. Since then, no one knows how to interpret the tablets. Several linguists have tried, but there is no generally accepted theory of how to read the symbols.[14]

There are no words to say, except for this one: *tragic*.

We end this chapter with the words of Ivan Petricevic, "the founder of *Curiosmos* and *Pyramidomania*, the latter being a project that aims to portray all pyramids on Earth in one giant map, with their respective information, images, and links."[15] He says:

I firmly believe that many of the megalithic sites we study today have not revealed all their secrets. Another fascinating example is Göbekli Tepe, an ancient site [close to the city of Şanlıurfa, and situated in Southeastern Anatolia, Turkey]. I really admire, not only because of what it signifies but because it is a perfect example of how advanced our ancestors were. Göbekli Tepe, for one, was built around 12,000 years ago. By history books, we are told that during this time, the region where Göbekli Tepe was built was inhabited by hunter-gatherer societies. The people that built Göbekli Tepe created something unprecedented; never before in history was such a massive site built, using such heavy stones.[16]

I use these words directly above since they are wholly appropriate for the situations that are found at Stonehenge, at Egypt, at Avebury, and, of course, at Easter Island.

CHAPTER 6

GIGANTIC!

In the introduction, I made a *very* brief reference to what is called Baalbek. Its role in this story, however, is incredible.

"Massive" would be far more appropriate. *World History* says that Baalbek

> *is an ancient Phoenician city located in what is now modern-day Lebanon, north of Beirut, in the Beqaa Valley. Inhabited as early as 9000 BCE, Baalbek grew into an important pilgrimage site in the ancient world for the worship of the sky-god Baal and his consort Astarte, the Queen of Heaven in Phoenician religion (the name "Baalbek" means Lord Baal of the Beqaa Valley). The center of the city was a grand temple dedicated to Astarte and Baal and the ruins of this early temple remain today beneath the later Roman Temple of Jupiter Baal. Baalbek is listed by UNESCO as a World Heritage Site.*[1]

Baalbek just happens to be home to an ancient quarry housing the so-called Stone of the Pregnant Woman. It's a totally massive stone block that weighs in at an incredible 1,000.12 tons. A second, huge block, exceeding the weight of the more famous "Stone of the Pregnant Woman" by roughly 240 tons, was found in the 1990s. Then, in 2014, a third huge, awesome stone was found. Its weight is estimated to be no less than 1,650 tons—making it the largest and the heaviest stone ever fashioned in recorded human history.[2] No wonder people flock to it. No wonder people come back shocked and awed. And, no doubt, with questions about how such blocks could have been created.

Controversy on TV

The History Channel's show *Ancient Aliens* notes that at this particular site "stand the ruins of Heliopolis built in the fourth century BC by Alexander the Great to honor Zeus. But beneath the Corinthian columns and remnants of both Greek and Roman architecture lie the ruins of a site that is much, much older. According to archaeologists it dates back nearly 9,000 years. It is the ancient city of Baalbek, named after the early Canaanite deity Baal."[3]

Ancient Aliens continued with these words: "Archaeological surveys have revealed that the enormous stone foundation that lies at the base of the site dates back tens of thousands of years, but even more significant to ancient astronaut theorists is their belief that the colossal stone platform may once have served as a landing pad for space travelers."[4]

The late Zechariah Sitchin, who was a devotee of the theory that our world was visited by extraterrestrials thousands of years ago, said of Baalbek:

> *The enigmas surrounding the site and the colossal stone blocks do not include one puzzle—where were those stone blocks quarried; because at a stone quarry about two miles away from the site, one of those 1,100-ton blocks is still there—its quarrying unfinished. . . . The quarry is in a valley, a couple of miles from the site of the "ruins." This means that in antiquity, someone had the capability and technology needed for quarrying, cutting and shaping colossal stone blocks in the quarry—then lifting the stone blocks up and carrying them to the*

construction site, and there not just let go and drop the stone block, but place them precisely in the designated course.[5]

Giant-sized stones at Baalbek (Wikimedia Commons)

The Most Famous, Ancient ETs, According to Sitchin

As for who just might have been able to achieve such a thing, Sitchin didn't have any worries about his theories and concepts. He was completely sure who got things going: "The great stone platform [at Baalbek] was indeed the first Landing Place of the Anunnaki gods on Earth, built by them before they established a proper spaceport. It was the only structure that had survived the Flood, and was used by Enki and Enlil as the post-Diluvial headquarters for the reconstruction of the devastated Earth."[6]

As fascinating as Sitchin's words concerning Baalbek and its origins certainly were, they amounted to collective theories. Hard facts, they were not. And they still aren't. Nevertheless, the fact is that the huge stones *are* there, for one and all to see. Also, there *are* multiple questions about how such massive, multi-ton stones could have been fashioned. So, whether the work of people or of the Levitators—or, even, a combination of both—there is still a mystery (or several or more) to be resolved. Another thought: What if the Levitators *were* the Anunnaki? Sitchin would likely turn all of this into hard, immovable fact. I can only provide you with theories surrounded by supportive data and worldwide, ancient structures.

Profiling the Anunnaki and More on the Mysteries of Baalbek

In 1999, the late Lawrence Gardner, author of such books as *Realm of the Ring Lords* and *Lost Secrets of the Sacred Ark*, said of the Anunnaki:

> *They were patrons and founders; they were teachers and justices; they were technologists and kingmakers. They were jointly and severally venerated as archons and masters, but [they] were certainly not idols of religious worship as the ritualistic gods of subsequent cultures became. In fact, the word which was eventually translated to become "worship" was avod, which meant quite simply, "work."*
>
> *The Anunnaki presence may baffle historians, their language may confuse linguists and their advanced techniques may bewilder scientists, but to dismiss them is foolish. The Sumerians have themselves told us precisely who the Anunnaki were, and neither history nor science can prove otherwise.*[7]

The *New Yorker* also has something to say about the enigma of Baalbek: "Nobody seems to know on whose orders it was cut, or why, or how it came to be abandoned. Baalbek is named for Baal, the Phoenician deity, although the Romans knew the site by its Greek name, Heliopolis. In short, Baalbek is a definitive puzzle. Also in the Baalbek complex is the Jupiter at Heliopolis. It, too, is incredible in size and design."[8]

Learn Religions has dug into the history surrounding the Jupiter at Heliopolis:

> *It is fitting that for the largest temple complex in the Roman Empire, Caesar would have the largest temples constructed. The Temple of Jupiter Baal ("Heliopolitan Zeus") itself was 290 feet long, 160 feet wide, and surrounded by 54 massive columns each of which were 7 feet in diameter and 70 feet tall. This made the Temple of Jupiter at Baalbek the same height as a 6-storey building, all cut from stone quarried nearby. Only six of these titanic columns remain standing, but even they are incredibly impressive.*[9]

I'll say they're impressive!

There's yet another notable angle that revolves around the Baalbek enigmas, too, as you'll see imminently. The *World Travel Guy* offers significant, important words on some of the history of Baalbek:

> *Over the years, the temple of Baalbek (aka Heliopolis) was damaged pretty extensively by earthquakes, and occasionally plundered for stone. At different times in history, it's been occupied by the Phoenicians, Greeks, Romans, Byzantines, Ottomans, and even the Mongols. Interestingly, some of the Corinthian columns were even disassembled by the Byzantine Empire and used in the 537 AD construction of the Hagia Sophia in Istanbul, Turkey, which is known today as one of the most famous wonders of the ancient world.*[10]

And, yet, these massive blocks still remain.

A Name Everyone Knows

Someone else recognized that the Baalbek stones amounted to such a famous, combined wonder. That man was none other than Mark Twain. It is well-worth noting that Twain, back in the 19th century, made a visit to Baalbek and, having done so, came back wholly amazed and shocked. Twain's very words alone—chronicled by his own hand—demonstrate the sheer amazement that built up within him when he reached Baalbek and saw those incredible stones.

Mark Twain put the following words down to posterity, something that was destined to ensure that future generations would be able to understand and appreciate the incredible nature of the Baalbek stones:

> *At eleven o'clock, our eyes fell upon the walls and columns of Baalbek, a noble ruin whose history is a sealed book. It has stood there for thousands of years, the wonder and admiration of travelers; but who built it, or when it was built, are questions that may never be answered. One thing is very sure, though. Such grandeur of design, and such grace of execution, as one sees in the temples of Baalbek, have not been equaled or even approached in any work of men's hands that has been built within twenty centuries past.*[11]

Twain was far from being done; he could barely tear himself away from the view directly before him. Most of us could probably understand and appreciate that. He put down his thoughts:

> *The great Temple of the Sun, the Temple of Jupiter, and several smaller temples, are clustered together in the midst of one of these miserable Syrian villages, and look strangely enough in such plebeian company. These temples are built upon massive substructions that might support a world, almost; the materials used are blocks of stone as large as an omnibus—very few, if any of them, are smaller than a carpenter's tool chest—and these*

substructions are traversed by tunnels of masonry through which a train of cars might pass. With such foundations as these, it is little wonder that Baalbek has lasted so long.[12]

"You Wonder Where These Monstrous Things Came From"

The man of the hour had much more to ponder upon, all of which demonstrates just how Twain found it so hard to believe what he was seeing. Yet, he *was* seeing it all. And it all practically blew him away. He wrote:

> *The Temple of the Sun is nearly three hundred feet long and one hundred and sixty feet wide. It had fifty-four columns around it, but only six are standing now the others lie broken at its base, a confused and picturesque heap. The six columns are their bases, Corinthian capitals and entablature—and six more shapely columns do not exist. The columns and the entablature together are ninety feet high—a prodigious altitude for shafts of stone to reach, truly—and yet one only thinks of their beauty and symmetry when looking at them; the pillars look slender and delicate, the entablature, with its elaborate sculpture, looks like rich stucco-work.*[13]

There's more from Twain; all of it made it clear he wasn't prepared for what he *did* see:

> *But when you have gazed aloft till your eyes are weary, you glance at the great fragments of pillars among which you are standing, and find that they are eight feet through; and with them lie beautiful capitals apparently as large as a small cottage; and also single slabs of stone, superbly sculptured, that are four or five feet thick, and would completely cover the floor of any ordinary parlor. You wonder where these monstrous things came from, and it takes*

some little time to satisfy yourself that the airy and graceful fabric that towers above your head is made up of their mates. It seems too preposterous.[14]

A Race of Gods or Giants

Undeniably amazed, Twain eventually got passed being stuck for words. His words soon tumbled out of his mouth, such was his incredibility for the view:

> *The Temple of Jupiter is a smaller ruin than the one I have been speaking of, and yet is immense. It is in a tolerable state of preservation. One row of nine columns stands almost uninjured. They are sixty-five feet high and support a sort of porch or roof, which connects them with the roof of the building. This porch-roof is composed of tremendous slabs of stone, which are so finely sculptured on the underside that the work looks like a fresco from below. One or two of these slabs had fallen, and again I wondered if the gigantic masses of carved stone that lay about me were no larger than those above my head. Within the temple, the ornamentation was elaborate and colossal. What a wonder of architectural beauty and grandeur this edifice must have been when it was new! And what a noble picture it and its statelier companion, with the chaos of mighty fragments scattered about them, yet makes in the moonlight!*[15]

Echoing the amazing and mysterious ways that the stones must have been fashioned, the time came when Twain simply couldn't keep his mouth shut:

> *I cannot conceive how those immense blocks of stone were ever hauled from the quarries, or how they were ever raised to the dizzy heights they occupy in the temples. And yet these sculptured*

blocks are trifles in size compared with the rough-hewn blocks that form the wide verandah or platform which surrounds the Great Temple. One stretch of that platform, two hundred feet long, is composed of blocks of stone as large, and some of them larger, than a street-car. They surmount a wall about ten or twelve feet high.[16]

The Giants of That Old Forgotten Time

"I thought those were large rocks, but they sank into insignificance compared with those which formed another section of the platform," said Twain. He continued onward:

These were three in number, and I thought that each of them was about as long as three street cars placed end to end, though of course they are a third wider and a third higher than a street car. Perhaps two railway freight cars of the largest pattern, placed end to end, might better represent their size. In combined length these three stones stretch nearly two hundred feet; they are thirteen feet square; two of them are sixty-four feet long each, and the third is sixty-nine.

They are built into the massive wall some twenty feet above the ground. They are there, but how they got there is the question. I have seen the hull of a steamboat that was smaller than one of those stones. All these great walls are as exact and shapely as the flimsy things we build of bricks in these days. A race of gods or of giants must have inhabited Baalbek many a century ago. Men like the men of our day could hardly rear such temples as these.[17]

"Some Great Ruin"

Now, we come to the final quote from Twain:

> *We went to the quarry from whence the stones of Baalbek were taken. It was about a quarter of a mile off, and downhill. In a great pit lay the mate of the largest stone in the ruins. It lay there just as the giants of that old forgotten time had left it when they were called hence—just as they had left it, to remain for thousands of years, an eloquent rebuke unto such as are prone to think slightingly of the men who lived before them. This enormous block lies there, squared and ready for the builders' hands—a solid mass fourteen feet by seventeen, and but a few inches less than seventy feet long! Two buggies could be driven abreast of each other, on its surface, from one end of it to the other, and leave room enough for a man or two to walk on either side.*
>
> *One might swear that all the John Smiths and George Wilkinsons, and all the other pitiful nobodies between Kingdom Come and Baalbek would inscribe their poor little names upon the walls of Baalbek's magnificent ruins, and would add the town, the county and the State they came from—and swearing thus, be infallibly correct. It is a pity some great ruin does not fall in and flatten out some of these reptiles, and scare their kind out of ever giving their names to fame upon any walls or monuments again, forever.*[18]

That we still have this amazing, old, journal-style book is incredible. Twain's emotive words still, to this day, make it very obvious that Baalbek is a place that is swamped in mystery—and *has* been swamped in mystery for thousands of years.

My Own Thoughts on Those Massive Stones at Baalbek

The thing that makes me conclude that the Baalbek stones were fashioned by something beyond us is (a) the incredible weight of those stones, and (b) the unlikely ability of the people of that era to have made and moved such stones that weighed in excess of one thousand tons. And, at Baalbek, there are no less than three of them, the most famous one being the "Stone of the Pregnant Woman." To put matters into fine perception, that particular stone alone is approximately the weight of three Boeing 747 Jumbo Jets.

When I say that I conclude the people of the area, and of the time, could not have created those three, massive slabs of stone at Baalbek—never mind move them—it's not meant as a slur on their undeniable skills when it comes to the world of stone. Nor does it make any issues regarding culture, countries, or skin color. To put it simply, my view is that, back then, *no one, anywhere, at all, on the planet* had the ability to create those three giant blocks and raise them into the sky. Unless, that is, you *were* one of the Levitators.

Atlas Obscura says of this issue: "This massive weight apparently proved too much for anyone to move, and the stone was left in the place where it was cut, an enormous rectangle sticking up at an angle from the ground. How Roman architects (or whoever it was that moved the trilithon) ever thought they would move such an enormous block remains a source of much debate."[19]

The most important issue in all of this, as I see it, is that the huge stones at Baalbek *were* meant to be used. Whether they were moved or not, is something else and very different. Clearly, the creators were sure they could have achieved their goal to get the massive stones on the move, so to speak. Otherwise, they wouldn't have gone ahead created them in the first place. We, today, wouldn't even try and sculpt such a thing. And certainly not three times over. But, for the Levitators it wasn't a big deal in the slightest.

CHAPTER 7

WHISTLING STONES INTO PLACE

Alicia McDermott writes of one particular place that cannot be ignored in a story like this one. She said:

> *The ancient Maya city of Uxmal is located in the Yucatan Peninsula in Mexico. It was one of the regional capitals during the Maya Late Classical period and is considered today to be one of the most important Maya archaeological sites. . . . Legend says that the first king of the city was bested by a magical dwarf, named Itzamna, who won Uxmal and the position as king by building the tallest monument (now known as the Pyramid of the Magician) in the city in one night.*[1]

UNESCO says the following of Uxmal and its history:

> *The Mayan town of Uxmal, in Yucatán, was founded c. A.D. 700 and had some 25,000 inhabitants. The layout of the buildings, which date from between 700 and 1000, reveals a knowledge of astronomy. The Pyramid of the Soothsayer, as the Spaniards called it, dominates the ceremonial center, which has well-designed buildings decorated*

with a profusion of symbolic motifs and sculptures depicting Chaac, the god of rain. The ceremonial sites of Uxmal, Kabah, Labna and Sayil are considered the high points of Mayan art and architecture.[2]

"Whistling" Stones to Where You Want Them

There is more to the story—much more. All of it is both incredible and sensational. That curious, small figure, Itzamna, found the power (or, probably more likely, was given the power) that ensured massive stones could be moved ridiculously and effortlessly into the air. He did so—so the ancient tales suggest, at least—by using nothing other than the power of a whistle. Of course, what do flutes do? They make a sound. Or, rather, we are seeing yet another perfect example of a wild story of the distortion of acoustic levitation to move stones of giant sizes. Do I think there could be another answer to the story? No, I do *not* think there could be an alternative theory. From all across the world we've heard tales that were digested and that had those same two components attached to them: stones in the skies and some form of sound. There are way more other examples, as will soon become apparent.

Morris Jessup, a UFO investigator of the 1950s, who died under very curious circumstances—and who also knew an incredible amount about levitation and massive stone creations—heard of the Uxmal stories, too, and chronicled everything in his diaries. Having traveled around much of South America, Mexico, Beliize, and Guatemala, he most assuredly knew his stuff.

I expanded on what Jessup had to say while I was lecturing at the Contact in the Desert conference in Joshua Tree, California, in 2015. I said to the audience:

Uxmal has a fascinating legend attached to it that fits in neatly with the material contained in this particular chapter. The history of the Mayans states that its origins date back to around 500 AD, with significant building having taken place in the 9th and 10th centuries.

By the 16th century, however, Uxmal was a dead place—its people having abandoned it in widespread fashion. There is, however, an eye-opening story that revolves around what is called the Pyramid of the Magician; also the Pyramid of the Soothsayer and the Pyramid of the Dwarf. It's a Mesoamerican step pyramid that sits in the heart of the old Mayan ruins and is surrounded by lush, green trees.

"An Absolute Minimum of Effort"

Also at Contact in the Desert, I said to the audience:

There is a notable reason as to why one of the several names of the pyramid is the Pyramid of the Dwarf. Legend has it that a mysterious, small humanoid—or, perhaps, humanoids—oversaw the construction of the 131-foot-high pyramid. In addition, the same dwarf was said to have been able to raise and direct the stones into place via a very novel and alternative fashion: namely, by whistling in the direction of the stones, something that caused them to rise up, and allowing them to be put into place with an absolute minimum of effort.

I continued:

We should not assume that one can move vast stones through the air just by whistling at them. Without a doubt that's utterly absurd. What is not so absurd, however, is how acoustic levitation can lift objects—and how the process involves sound. It's entirely plausible that the Mayans knew that sound played a role in the stone-raising. And, as a result, and over a fairly significant period of time, turned the science of acoustic levitation into simplistic whistling in their legends and folklore.

CHAPTER 8

STRANGENESS IN THE UK

Bringing up the matter of near-magical, large stones in the heart of the United Kingdom is almost bound to provoke images in the minds of so many people of legendary Stonehenge. But hang on just a moment. The amazing construction is far from being alone in the UK. Indeed, there are way more than just a few such creations dotted all across the countryside. As you will soon see, in so many of the cases under the microscope, there's a fascinating theme ever-present in the country: Namely, it's that of ancient, large stones that, centuries ago, *moved. Here, there, and everywhere.* That sounds bizarre, I know. And, seemingly, according to the old tales and legends, they did so *of their very own volition.* And let their weights and their sizes be damned. We begin with the most appropriately titled "Dancing Stones" of Stackpole, Pembrokeshire, Wales, a legendary trio of stones that are spread roughly a mile from each other—and that, very eerily, are apparently incapable of standing still. Or, in the old tales, that is how things were for many centuries.

As the ancient legend goes, once a year the three stones meet up at on the old, Welsh hills and fields. Slowly, and by nothing but an atmospheric, bright moonlight, they eerily begin to move. Slowly, bit by bit. Then, that movement begins to become more and more animated. And, finally, the three then start to engage in what can only be described as a full-blown dance—right up until dawn breaks and weirdness is gone. What is particularly intriguing about this story is that, as it's revealed, the stones would dance to the *sound of music.* In this tale, the noise is provided, in the form of a flute, by none other than the Devil himself. So, yet again, we have an ancient story—circulated over the decades, and certainly changed over time—of stones that were magically moved by the mysterious medium of sound. And, there's the presence of a nonhuman entity; in this particular case, it is said to be in the form of the Devil. Perhaps, though, it was really one of those enigmatic, but amazing, Levitators of old. Ponder on that.

"Some Unknown Power"

Janet and Colin Bord, both authors and experts on the ancient history of the United Kingdom, have made careful studies of these legends of walking stones that can be found across much of the United Kingdom. The husband and wife team say, for example, something very strange that's tied to stones that are known to move: "On Midsummer Eve, the stones of St. Lythans [situated in southeast Wales] chambered cairn near St. Nicholas (South Glamorgan) whirl around three times and curtsey. The stones of the Grey Wethers stone circle on Sittaford Tor, Dartmoor (Devon), go for a short walk at sunrise, while the Longstone, above Chagford (Devon), turns round slowly at sunrise in order to warm each side in turn."[1]

Another fascinating story from the Bords' voluminous archives goes as follows: "The Wergin Stone, Sutton, near Hereford was moved 240 paces from its former position some time in 1652, no-one knew how, and it needed nine yoke of oxen to take it back."[2]

The Bords have yet another account to tell, from the Republic of Ireland, from my visit in 1998: In the ancient graveyard stand the remains of St. Fechin's church, partly built. The huge lintel stone over the doorway weighs more than two and a half tons, and legend has it that workmen were unable to lift this stone into position. Such a feat was not beyond St. Fechin, who used some unknown power (levitation?) to raise the stone. Other miracles attributed to him included making a nearby stream run uphill.

Stones and Witches

Fellow anomalies researcher Neil Arnold is someone who has done a lot of research into what is known as Kit's Coty House. It's an old stone construction located in the ancient county of Kent, England. Arnold, who has spent a lot of time at the stones, described them to me as

> *a set of Neolithic stones said to be older than Stonehenge," which jut from a field at Blue Bell Hill. These stones have a lot of folklore attached to them. Some suggest that the stones are used as a calendar or could be a mark of where a great and bloody battle once took place. Others believe the stones to have once been used for sacrificial means, and there are those who opt for the more fanciful rumour that they were constructed by witches on a dark and stormy night.*[3]

The enigma of the spinning stones (Wikimedia Commons)

Very important to the story above is the legend that says not only were the stones the work of a group of old witches, but that those same witches could command the stones to move, and to move them just about anywhere.

English Heritage says of the site: "The origin of the name Kit's Coty is not known. For many years it was thought to be a corruption of Catigern, the name of a British prince slain in single combat with the Saxon Horsa in a battle at Aylesford in AD 455 at which the Britons were victorious. The monuments were therefore assumed to be a memorial to him."[4]

Whirling Wildly

Our foray around the United Kingdom is hardly over. In August 2000, a man named David Farrant, a now-late, self-described vampire hunter, and the author of *The Vampyre Syndrome*, *Shadows in the Night*, and *Beyond the Highgate Vampire,* claimed to have seen, late one night, the stones that comprise Kit's Coty House, spinning at an absolutely furious speed, but without affecting or damaging the formation in any way at all. And, with not even a solitary bit of damage to the surrounding grass when the spinning was over, either.

Farrant said that on that appropriate dark, August night, he was hanging out at Kit's Coty House to perform a midnight ritual that was designed to bestow wealth on he or she who cracks the secrets of the moving stones of Kit's Coty House.[5] Farrant—in a fascinating, but cloak and dagger type fashion—claimed that he had cracked the secrets of moving the stones of the County of Kent. He said the answer was (in one, single word) magnets. Rather frustratingly, that's *all* Farrant would say. However, the strange issue of magnets raising heavy and large stones will crop up later—and in an equally fascinating story. This time, the location will be Florida, in the United States of America, and the key figure will be a man named Edward Leedskalnin: a man who possibly solved the secrets of Egyptian antigravity—but chose to keep it all to himself, and whose strange story entranced none other than Billy Idol.

Held in Considerable Awe

Born in 1853, Marie Trevelyan was the author of a well-revered book, *Folk-Lore and Folk-Stories of Wales*. She chronicled one particularly relevant story that involved sound and a stone—but this time in an undeniably weird fashion. Arguably, it's even weirder than the tale that David Farrant told. It all revolves around what became known, centuries ago, as the Deity Stone. It's located in the town of Penmaenmawr, Conwy County, North Wales. Trevelyan wrote:

> *The Druids' Circle, which is about a mile distant from the Green Gorge on Penmaenmawr, contains two stones among other Druidical remains. The Deity Stone was formerly held in considerable awe. An old story told by a North Welshman was to the effect that if anybody used profane language near it, the stone would bend its head and smite the offending person. A man from South Wales played cards with some friends beside this stone on a Sunday, and when the men returned to the village with cuts about their heads, the people knew the Deity Stone had smitten them, though they would not admit having had punishment.*[6]

This particular tale isn't quite over yet.

When a Stone Becomes Violent

"A notorious blasphemer who came from Merionethshire laughed to scorn the story of this stone," said Trevelyan. "One night he went to the Druids' Circle alone and at a very late hour, and shouted words of blasphemy so loud that his voice could be heard ringing down the Green Gorge. People shuddered as they heard him. The sounds ceased, and the listeners ran away in sheer fright. In the morning the blasphemer's corpse was found in a terribly battered condition at the base of the Deity Stone."[7]

There were shouts, there were sounds, and there were what seemed to have been a full-on "attack" on the man, suggesting that the stone moved—and moved violently, no less. Yet again, as I see it, we are looking at an extremely warped version of sound being utilized to try and move the Deity Stone. A situation that went drastically wrong during an operation involving sound and stone? Don't bet against that, at all. That just might have been *exactly* what happened all those years ago.

"Dreams of a Great Pile of Stones"

Now, onto the strange story of the curiously titled "Cheesewring." It has nothing to do with cheese in the slightest. Rather it's a form of ancient granite and it's located on the huge, wild, 80-square-mile Bodmin Moor, England. Should you wonder what the moor looks like, I'll tell you. Imagine a huge, expansive slab of landscape that is both inviting and atmospheric during the day—aside from when it isn't swamped in creepy, dense fog—and is chilly with howling winds at night. Starlit skies are typical. The adjacent Dartmoor was the site of Sir Arthur Conan Doyle's classic Sherlock Holmes 1902 story, *The Hound of the Baskervilles*. That may give you some idea of just how atmospheric that ancient moor really is.

Wilkie Collins, a 19th-century author and playwright whose many works included *The Woman in White*, *The Dead Secret*, and *The Moonstone*, said of the Cheesewring: "If a man dreams of a great pile of stones in a nightmare, he would dream of such a pile as the Cheesewring. All the heaviest and largest of the seven thick slabs of which it is composed are at the top; all the lightest and smallest at the bottom. It rises perpendicularly to a height of thirty-two feet, without lateral support of any kind."[8]

"Distortions of the Old Stories"

Collins continued with his story of the Cheesewring: "The fifth and sixth rocks are of immense size and thickness, and overhang fearfully all round the four lower rocks which support them. All are perfectly irregular; the projections of one do not fit into the interstices of another; they are heaped up loosely in their extraordinary top-heavy form on slanting ground, half way down a steep hill."[9]

Then, there is the legend that whenever a crow flies over the Cheesewring—specifically while it's crowing loudly—the Cheesewring granite itself will turn around three times. Once again we have two key issues involved: (a) sound, in the form of a crow, and (b) imposing stones moving at the very same time. This particular chapter, perhaps more than any other one, demonstrates that the legends of ancient stones and rocks that magically moved across landscapes here and there—if not *everywhere*, too—were spread both far and wide. They were nowhere near restricted to Egypt, Baalbek, South America, Easter Island, and Central America. These stories popped up in tiny, ancient English villages.

One more thing for this chapter: It's well worth noting that Mog Ruith, a famous figure in ancient Irish mythology, supposedly had the ability to turn people into stone. Notably, when transformed into stone, those same people would still able to move. Yet again, distortions of the old stories of acoustic levitation and of stones moved above the ground? What else could it be? Nothing, that's what.

CHAPTER 9

HANGING OUT WITH THE ROLLING STONES . . . KIND OF

Situated in an inviting area of rural Oxfordshire, England, is a formation of ancient stones known as the Rollright Stones. They sit near an equally ancient village called Long Compton and they are surrounded by lush, green land. On the official website of the legendary formation, there are the following words: "This complex of megalithic monuments lies on the boundary between Oxfordshire and Warwickshire, on the edge of the Cotswold hills. They span nearly 2,000 years of Neolithic and Bronze Age development and each site dates from a different period."[1]

We're also told of these particular stones:

> *The oldest, the Whispering Knights dolmen, is early Neolithic, circa 3,800-3,500 BC, the King's Men stone circle is late Neolithic, circa 2,500 BC; and the King Stone is early to middle Bronze Age, circa 1,500 BC. The Stones are made of natural boulders of Jurassic oolitic limestone which forms the bulk of the Cotswold hills. This stone has been used extensively in the region*

for building everything from churches and houses to stone walls. The boulders used to construct the Rollright Stones were probably collected from within 500m of the site.[2]

Some Very Strange Things about the Rollright Stones

From *Britain Express*, we have this:

> *The stones are irregularly spaced, but drawings made by antiquarians suggest that there was originally a continuous wall of 105 stones placed side by side, with a small opening in the south west. Now only 70 stones remain, with several obvious gaps in the circle. Two stones, one standing, one fallen, lie outside the circle, flanking the opening, which is directly opposite the tallest stone and would have formed an entrance portal.*[3]

Just like so many other ancient structures within the United Kingdom, there are some deeply weird things about the Rollright Stones, as you will now see. Back in the early years of the 17th century, one William Camden, a historian, became fascinated—*obsessed* would be *way* more accurate—with the Rollright Stones. He believed that they were really those protective aforementioned knights, but morphed into stone by a crone-like, old soothsayer and prophetess who went by the moniker of Mother Shipton. And just who might she have been? Read on and you'll soon become acclimatized to this sinister character from centuries ago:

> *Mother Shipton was born Ursula Sontheil in 1488, during the reign of Henry VII, father of Henry VIII. Although little is known about her parents, legend has it that she was born during a violent thunderstorm in a cave on the banks of the River Nidd in Knaresborough. Her mother, Agatha, was just fifteen years old*

when she gave birth, and despite being dragged before the local magistrate, she would not reveal who the father was.

With no family and no friends to support her, Agatha raised Ursula in the cave on her own for two years before the Abbott of Beverley took pity on them and a local family took Ursula in. Agatha was taken to a nunnery far away, where she died some years later. She never saw her daughter again.[4]

Strange phenomena at a UK stone circle (Wikimedia Commons)

It's now the time to come to the most important, and relevant, part of this uncanny story. It mirrors so much of what we have seen so far—namely, the strange, and the near-unfathomable, movements of stones under incredible ways and means.

"One Man's Acoustic Levitation Was Another Man's Magic Stones"

In the latter part of the 1970s, a UK-based team of people, with a deep fascination for the history and folklore that surrounds the Rollright Stones, created what they termed the Dragon Project. it was an undeniably ambitious program that was designed to resolve just about all of the secrets of the ancient stones of the United Kingdom. One of the key figures in the operation was a man by the name of Don Robins. He was based in Middlesex, England, trained as a chemist, completed a PhD in magnetism, and became a lecturer in applied science. In his 1985 book on the many and varied experiences of the team and more, titled *Circles of Silence*, Robins said something notable, and something that was highly relevant to what we've seen, so far, in relation to the theme of this book:

> *Some kind of rationalization of the petrified legends can be made, even though there are many loose ends and probably no direct link with the builders of the circles and their beliefs. Other legends are not so tame.* Underlying many of the petrification legends is a persistent theme that on certain days the stones move. The Rollright Stones walk to a nearby stream to drink, and other stones rotate on their axes *[emphasis mine].*[5]

See what I mean? There are stories like this one all across the United Kingdom.

As with what we know about the stones of Egypt, of Easter Island, and of Baalbek, we should most definitely not take the strange legends of walking stones literally, or, on the other hand, as 100-percent kosher. What we *should* consider, however, is that in ancient times, and when the Rollright Stones were created, acoustic levitation played the prime, key, decisive role in the construction *and also in the movement* of those very same stones. Not only that, it would have been very easy, in times long gone, for a science that was based on levitation and sound, and that eventually became forgotten, to have been spectacularly misinterpreted very quickly to a strange form of magic—and then finally, and largely, lost. In other words, many centuries ago, one man's acoustic levitation was another man's occult stones. Take your pick.

A Strange Creature on the Loose

There is another most important issue that adds strange weight to the equally strange story of the Rollright Stones. In a later chapter, we will see how acoustics, and particularly what is termed infrasound, can create graphic and, sometimes, terrifying, images of hallucinatory forms. Usually, they are of monstrous forms, and the worst type possible: hairy "ape-men" and werewolf-like monstrosities and sometimes even lake monsters of the "Loch Ness Monster" type. And, in relation to all of that, we'll also see how there is a curious and creepy connection among infrasound, the mysterious and ancient Levitators, acoustic levitation, antigravity, and strange, hideous creatures that prowl around ancient sites as terrifying "paranormal guard dogs."

The reason why, at this *very* point, I mention all of the above is because while the Dragon Project was deep into its research, in 1977, a strange, large, hair-covered creature was seen prowling around the Rollright Stones. Paul Devereux of the Dragon Project and still active in the field today, said that the witness was a highly respected archaeologist, and the man was adamant that he saw a huge, gray-haired-covered beast walk by.[6] In seconds, the thing was gone.

Monsters, walking stones, and levitation? I assure you there is method to all of the madness. One final thing on the really weird aspects of the Rollright Stones, we have the words of a man named Murry Hope. He is the author of many books, including *Practical Celtic Magic, Ancient Egypt: The Sirius Connection,* and *The Spell Book for New Witches.* He states that, after he took a visit to the Rollright Stones, it was as if something appeared to be "programmed to respond (or resonate with) a specific sonic, which had nothing to do with time as we know it. The original program must have been inserted either by an advanced race from the far past whose knowledge of sonics far exceeded the bounds of present-day science, or from an extraterrestrial source."[7]

What all of this demonstrates is that sound, in varying ways, has played a major role at the Rollright Stones for close to two thousand years. Later, we'll be immersed in those other forms of sound—all of them having a link to the Levitators—something that will demonstrate the incredible Levitators were driven by varying types of sound and acoustics.

CHAPTER 10

THE MYSTERIES OF STONEHENGE

Some ancient structures are more well-known than others. There's no doubt, however, that Stonehenge is very near the top of the list. Not only is it an undeniably magical and mysterious place, but it also plays a significant role in the overall story that this book tells. That's right: Those mighty stones may very well have been erected with the use of acoustic levitation. Arguably, that's the only way such an immense project could have been successfully achieved. Forget rollers. Forget ropes. Forget pulleys. Before we get to the matter of how Stonehenge became the iconic structure that it surely is today, let's see what we know about the place and its history.

I should stress that, although we have a great deal of data on the legendary creation itself, much of the story is still very much shrouded in enigmas, folklore, and myth. Stonehenge was, most archaeologists suggest, constructed roughly about 3100 BC. In other words, about five thousand years ago. That's not the full story, though: There is clear evidence that the area on which Stonehenge stands now—the English county of Wiltshire—was used by the people of an even earlier era, possibly even to as far back as 8000 BC.[1]

The Purpose of These Pits Is Still a Mystery

As for when Stonehenge was first studied to a fairly significant degree, the *English Heritage* comes to the point: "The first known excavation at Stonehenge, in the center of the monument, was undertaken in the 1620s by the Duke of Buckingham, prompted by a visit by King James I. The king subsequently commissioned the architect Inigo Jones to conduct a survey and study of the monument. Jones argued that Stonehenge was built by the Romans."[2] Jones was way off course.

Now let's put Stonehenge in its proportions. Literally. The giant formation is comprised of several different types of stone, not just one. And that's a very important issue. You'll see precisely why. We'll begin with what are termed the Aubrey Stones. *History in Numbers* says:

> *The Aubrey Holes are a series of chalk pits that circle the inner bank of Stonehenge. They are named after John Aubrey, the seventeenth-century biographer and philosopher who first observed and noted them. The purpose of these pits is still a mystery, with a common theory being that they were used to support stones or posts. Later, they appear to have been used to bury cremated remains.*[3]

Stonehenge: one of the world's most famous circles of stone (Nick Redfern)

A bank and a ditch are present, as are, of course, those mighty stones that are termed the Henge. And *mighty* is a most apt term for us to use. Like so many other enigmatic creations around the world, the stones of Stonehenge run to multi-tons in weight. For example, there are the bluestones, as they are known, some of which weigh in at about four tons. There are no fewer than 82 of those hulking, giant things.

Structures for Sound Amplification

Now, let's take a look at the most important part of the Stonehenge story: Why was it constructed? What was its purpose? The theories are many, something that inevitably makes things both complicated and controversial. *History* offers what is deemed by many to be a plausible theory:

> *One enduring hypothesis for Stonehenge's purpose comes from the initial observation, first made by 18th-century scholars, that the monument's entrance faces the rising sun on the day of the summer solstice. For many, this orientation suggests that ancient astronomers may have used Stonehenge as a kind of solar calendar to track the movement of the sun and moon and mark the changing seasons.*[4]

LiveScience has addressed this taxing, yet intriguing, issue, too. They took a look at the work of Rosemary Hill, the author of *Stonehenge: Wonders of the World,* a very good book on the legendary creation of the huge, stone circle. *LiveScience* said: "Researchers have proposed myriad ideas over the years, suggesting that monuments like Stonehenge were used as sacred hunting grounds, places of community gathering, astronomical calendars, structures for *sound amplification* [italics mine], cemeteries or even havens for ancient healing. Excavations offer supporting evidence for some of these claims."[5]

The Bluestones Didn't Come from Anywhere Near where Stonehenge Stands

"They've found [human] remains at Stonehenge, so that's strong evidence it was a burial site and it's orientated to the sunset during the winter solstice," Rosemary Hill explained. "So I think you can say it's to do with the dead and the solstices. It's not unreasonable to think of it as a ritual site and there's no evidence of people eating or living there."[6] The matter of sound amplification is an intriguing one, and an issue we'll return to later. Of course, there's a possibility that Stonehenge was created for multiple reasons, rather than just one.

Now, it's time to get to the most intriguing part of the story—and the history of Stonehenge, too. It's the curious matter of the movement of those stones that led to its construction. It's important to note that two primarily and distinctly different stones were used in the construction of Stonehenge: the sarsen stones and the bluestones. *English Heritage* says of the former:

> *For many years most archaeologists believed that these stones were brought from the Marlborough Downs, 20 miles (32km) away, but their exact origin remained a mystery. However, recent research using a novel geochemical approach has not only confirmed that the Marlborough Downs were indeed the source, but has pinpointed the specific area that the sarsens most likely came from—the area known as West Woods, south-west of Marlborough.*[7]

Taken from the Preseli Mountains

As for the specific weights of the sarsen stones, some of them tip the scales at about an astounding 25 tons. Impressive, right? It is, however, the latter stones—the bluestones—that are the most thought-provoking of all. You'll now see why. What is particularly eye-opening is that the bluestones didn't actually come from anywhere near where Stonehenge stands. They didn't even come from the county of Wiltshire, in which Stonehenge is situated today. In fact, their origins were not even in England. Somehow, those gigantic, super-heavy stones were brought all the way to Wiltshire, England, from no less than Wales. The distance from one to the other? Check this out: no less than approximately *240 miles.* Bear in mind that all those thousands of years ago there were no highways with multi-lanes of the type that we, today, take for granted. You want even more controversy? You've got it, for sure.

That particular type of stone came from Wales's Preseli Mountains. Yes, that stone had to be brought across *mountains. FelinFach* provides us with a good look at the mountains, themselves:

> *The Preseli Hills, or as many locals believe, the Preseli Mountains are a range of hills in west Wales that are in the western part of the U.K. and form part of the Pembrokeshire Coast National Park. The range stretches from Carn Ingli at Newport in Pembrokeshire in the west to Frenni at Crymych in the east, a distance of some 13 miles or 20 km. The river Nyfer begins its journey near to Crymych and flows its eleven miles almost entirely through the Pembrokeshire National Park and has its estuary in Newport Pembrokeshire. Other rivers whose source is in the Preseli Mountains include the Gwaun, Syfynwy and Taf.*[8]

Geoffrey of Monmouth and the Giants' Dance

Some might say that it is one thing to move massive amounts of stone across relatively flat landscapes—slopes, even—but it's a totally different thing when we are talking about negotiating *mountains.* And I should also stress that the Bluestones of Wales typically weighed about four tons. Were all of those stones taken through wild, mountainous landscapes close to 240 miles on nothing but timber rollers? That's one theory. By the way, it's a ridiculous theory and should be banished from the history books. The idea these massive stones were dragged by nothing but brute force has been addressed. That's insane, too. Toss it away. So, what's the answer to all of this? Well, it could very well be in the world of acoustic levitation. It is, in this writer's estimation, that Stonehenge was the work of the Levitators. Now, let's look at some other mysteries surrounding Stonehenge.

The Cambridge University Press says the following:

> *In the oldest story of Stonehenge's origins, the* History of the Kings of Britain *(c. AD 1136), Geoffrey of Monmouth describes how the monument was built using stones from the Giants' Dance stone circle in Ireland. Located on legendary Mount Killaraus, the circle was dismantled by Merlin and shipped to Amesbury on Salisbury Plain by a force of 15,000 men, who had defeated the Irish and captured the stones. According to the legend, Stonehenge was built to commemorate the death of Britons who were treacherously killed by Saxons during peace talks at Amesbury. Merlin wanted the stones of the Giants' Dance for their magical, healing properties.*[9]

While much of Geoffrey's story is, one can clearly tell, nothing but wild folklore and legend, it's most important to note that, as we've seen time and again, myths are very often born out of distorted realities that are often lost to the fog of time. In light of all that, it's well worth noting that in Geoffrey's tale—and in relation to the Giants and their dancing to music, too—the massive stones known today as Stonehenge were moved via a form of music directed by mysterious giants. Music that was really acoustic levitation? There's no reason to think that's not the answer. But there's *every* reason to think it *is* the answer.

The Enigmas of the Famous Stone Circle

Author of *Great Cities of the Ancient World*, L. Sprague de Camp, made it abundantly clear that there were some very strange stories surrounding the history of Stonehenge. He quoted our friend (our *very old* friend) Geoffrey of Monmouth, who said, all of those centuries ago:

> *If thou be fain to grace the burial-place of these men with a work that shall endure forever, send for the Dance of the Giants that is in Killiaranus, a mountain in Ireland. For a structure of stones is there that none of this age could arise save his wit were strong enough to carry his art. For the stones be big, nor is there stone anywhere of more virtue, and, so they be set up round this plot in a circle, even as they be now there set up, here shall they stand forever.*
>
> *For in these stones is a mystery, and a healing virtue against many ailments. Giants of old did carry them from the furthest ends of Africa and did set them up in Ireland what time they did inhabit therein. And unto this end they did it, that they might make them baths therein whatever they ailed of any malady, for they did wash the stones and pour forth the water into the baths, whereby they that were sick were made whole.*[10]

"A Sonic Property"

This chapter ends with something that is well worth scrutinizing. In 2014, the BBC revealed a story that has a connection to nothing less than acoustics and Stonehenge, itself. The title of that article? For us, it's a very important title: "Stonehenge Bluestones Had Acoustic Properties, Study Shows."

The BBC said of this: "The giant bluestones of Stonehenge may have been chosen because of their acoustic properties, claim researchers. A study shows rocks in the Preseli Hills, *the Pembrokeshire source of part of the monument, have a sonic property* [italics mine]. Writer Paul Devereux [of the *Dragon Project* of the 1970s] said: 'It hasn't been considered until now that sound might have been a factor.'"[11]

"The Primeval Equivalent of Lourdes"

In 2008, the *Seattle Times* newspaper highlighted the means by which Stonehenge could heal the sick. They wrote:

> *The first excavation of Stonehenge in more than 40 years has uncovered evidence that the stone circle drew ailing pilgrims from around Europe for what they believed to be its healing properties, archaeologists said Monday. Archaeologists Geoffrey Wainwright and Timothy Darvill said the content of graves scattered around the monument and the ancient chipping of its rocks to produce amulets indicated Stonehenge was the primeval equivalent of Lourdes, the French shrine venerated for its supposed ability to cure the sick. An unusual number of skeletons recovered from the area showed signs of serious disease or injury. Analysis of their teeth showed about half were from outside the area.*[12]

"Archaeologists Have Finally Realized that Ancient People Had Ears"

We've seen how, to some degree, acoustics played a role in eras past, and particularly relative to human health. To show that those who came before us knew *exactly* what they were saying, take note of these words from *Scientific American*:

> *Kullervo Hynynen, a medical physicist at Sunnybrook Research Institute in Toronto, and a team of physicians are trying out a technique that involves giving patients a drug followed by an injection of microscopic gas-filled bubbles. Next patients don a cap that directs sound waves to specific brain locations, an approach called high-intensity focused ultrasound. The waves cause the bubbles to vibrate, temporarily forcing apart the cells of the blood-brain barrier and allowing the medication to infiltrate the brain.*[13]

Let's get back to Paul Devereux. He said, in 2011, something vitally important. So important were his words, I'll place the whole sentence in eye-catching italics: "*Archaeologists have finally realized that ancient people had ears, and have discovered that various kinds of acoustic effects—from eerie echoes to resonant frequencies that can affect the brain—seem to have been an intentionally planned component of a number of prehistoric sites worldwide, from ruined temples to rock art locations.*"[14]

"It's Activating the Normal Healing Process"

These words of Devereux deeply intrigued the BBC (and certainly it's no wonder): "A blast of ultrasound can help stubborn chronic wounds heal more quickly, a study suggests."[15] Tests on animals, published in the *Journal of Investigative Dermatology*, showed healing times could be cut by nearly a third.[16]

There was more to come from this then-breaking news: "It's activating the normal healing process, that's why it's an attractive therapy; the ultrasound is simply waking up cells to do what they do normally. The researchers now need to study the approach in people, which they expect to do in the next year. 'We're looking at 200,000 patients currently with a chronic wound, all those may well benefit from the technology,' Dr. Bass said."[17]

Sounds and stones: we are, bit by bit, seeing just how they are so interlocked in this story—and how that has been the case for thousands of years. The Levitators knew so much of the importance of all that and its relevance to their lives and society.

CHAPTER 11

USING SOUND FOR GOOD AND BAD

So far, we've seen acoustic levitation—thousands of years ago—primarily utilized to construct buildings of massive proportions and with incredible ease. In this chapter, however, things become very different. One part of the story concerns the destruction of a city. The other section revolves around the parting of the Red Sea. We'll see how, in the past, buildings were completely razed to the ground by using directed sound waves and how, centuries ago, it was used in a totally malevolent fashion.

"The Red Sea, one of the most saline bodies of water in the world, is an inlet of the Indian Ocean between Africa and Asia," say those who run the *New World Encyclopedia*. They add: "The connection to the ocean is in the south through the Bab el Mandeb sound and the Gulf of Aden. In the north are the Sinai Peninsula, the Gulf of Aqaba, and the Gulf of Suez (leading to the Suez Canal). The Sea has played a crucial navigational role since ancient times."[1]

Got Questions says of the legendary parting of the sea, as described in the pages of the Old Testament:

> *The importance of the parting of the Red Sea is that this one event is the final act in God's delivering His people from slavery in*

Egypt. The exodus from Egypt and the parting of the Red Sea is the single greatest act of salvation in the Old Testament, and it is continually recalled to represent God's saving power. The events of the exodus, including the parting and crossing of the Red Sea, are immortalized in the Psalms as Israel brings to remembrance God's saving works in their worship.[2]

Acoustic levitation and parting the Red Sea (Wikimedia Commons)

There are, however, conventional, and down-to-earth theories for what promoted the tale of the parting of the Red Sea. There's a distinct possibility that the legendary body of water was parted by sound. Before we get to that theory, there's another angle. *Smithsonian Magazine* has highlighted the work of software engineer Carl Drews. Rachel Newer, a writer for the magazine, says: "Given the conditions of the lake a couple thousand years ago, a coastal phenomenon called a 'wind setdown'—very strong winds, in other words—could have blown in from the east, pushing the water to create a storm surge in another part of the lake, but completely clearing water from the area where the wind was blowing."[3]

Legend or Something More?

Plausible? Sure. But, what of antigravity? Is it possible that acoustic levitation was the real cause of what it was that parted the Red Sea? These are the questions that should be addressed for one particular reason: Right now, research is being undertaken to see how water just might be manipulated by levitation, even though admittedly still in small amounts.

On this issue, we have these words from the BBC:

> *this requires extremely powerful sound to levitate even a tiny droplet of water, so the researchers have worked with ultrasound—frequencies beyond those that human ears can detect—to avoid damaging their hearing. Researcher Daniele Foresti explained that, although acoustic levitators were not a new concept, this was the first one able to move and control the material it levitated. 'We have total control of the acoustic field inside,' he told BBC News.*[4]

This tells us that if, one day, we do manage to levitate, and "move" water in huge amounts, it will undoubtedly take a long time. Basically, we're way behind what the Levitators succeeded in doing so many years ago. That's not to say we won't manage to achieve such an awe-filled situation. It's just going to take a while. Or longer, even. Let us stay positive, though. Those legendary, famous words *Mighty oaks from little acorns grow* immediately spring to mind. Let's keep them in *our* minds.

"It's Real"

There are those in the field of conspiracy theorizing who suspect that, to some degree, at least some of this may have been achieved. In a 2019 article for Mysterious Universe titled "Commanding and Controlling the Weather by 2025?" I wrote:

In April 1997, William S. Cohen, who was then the U.S. Secretary of Defense when Bill Clinton held the position of President of the United States, made an amazing and controversial statement to a packed audience at the University of Georgia, which is located in Athens. The conference at which the Secretary of Defense was speaking was The Conference on Terrorism, Weapons of Mass Destruction, and U.S. Strategy.[5]

Cohen came straight to the point and told the audience that certain agencies and people around the world—whom he tactfully elected not to name, which is intriguing—were then presently "engaging in an eco-type of terrorism whereby they can alter the climate, set off earthquakes, volcanoes remotely through the use of Electro-Magnetic waves. So there are plenty of ingenious minds out there that are at work finding ways in which they can wreak terror upon other nations. It's real."[6]

Cohen's words came 25 years before I wrote this article. Where might things be now? Who knows? If a U.S. Secretary of Defense can come forward and confirm on the record that weather, as far back as 1995, was in the process of being weaponized, and on a large scale, then it's highly plausible that acoustic levitation and weather combined—whether for war or for good purposes—were used to bring down the Walls of Jericho and to part the Red Sea.

Bringing Down a City in a Decidedly Alternative Fashion

The legendary Ark of the Covenant, made famous in Steven Spielberg's hit 1981 movie, *Indiana Jones and the Raiders of the Lost Ark,* now comes to the surface. The story demonstrates the incredible, destructive power the ancients had in their hands all those millennia ago. Before we get to the matter of destruction itself, let's see what led up to the events that saw the famous walls of Jericho come tumbling down. What we know from the Book of Joshua, in the pages of the Old Testament, is that the city of Jericho existed in what is now the West Bank, surrounded by the Dead Sea and Jordan to the east), and by Israel to north, south, and west. *Britannica* offers this description of the early history of the location:

> *Jericho is one of the earliest continuous settlements in the world, dating perhaps from about 9,000 bce. Archaeological excavations have demonstrated Jericho's lengthy history. The city's site is of great archaeological importance; it provides evidence of the first development of permanent settlements and thus of the first steps toward civilization. Traces have been found of visits of Mesolithic hunters, carbon-dated to about 9,000 bce, and of a long period of settlement by their descendants.*[7]

High-tech weaponry in ancient times (Wikimedia Commons)

"At Joshua's Order, the Men Produced a Powerful Roar"

For our purposes, we're going to keep our sights on the Old Testament. Bible Study Tools states how the Battle of Jericho began:

> *The Bible Story of the Battle of Jericho is found in the book of Joshua, chapter six. This was the first conquest of the Israelites as they set out to win the land of Canaan [and] God instructed Joshua with an unusual strategy for the battle of Jericho. He told Joshua to have his army march around the city once a day for six straight days. While marching, the soldiers played their trumpets as the priests carried the Ark of the Covenant around the city of Jericho. At Joshua's order, the men produced a powerful roar, and* Jericho's walls miraculously fell down [emphasis mine]. *The Israelite army raced in quickly conquering the city and, as promised, only Rahab and her family were spared.*[8]

Let's look closer at the Ark of the Covenant and its awesome powers. Matters are made somewhat confusing by the claims that the Ark was clearly different things to different people. For some, it was a means by which one could speak with no less than God, himself. For others, it was a device that could create incredible destruction. Interestingly, some researchers have suggested that the Ark was dangerously radioactive, although that's particularly controversial.

The City's Huge, Stone Walls Brought to the Ground by Trumpets

On top of that, there's a body of data that suggests the Ark of the Covenant could kill with electricity. It seems unlikely that such a thing could exist—and have such multiple powers—and yet, that is precisely what the Old Testament tells us. What's also intriguing is that the Ark was hardly impressive in terms of its size. Writer Owen Jarus gives us this: "The story of the construction of the ark told in the *Book of Exodus* describes in great detail how God ordered Moses to tell the Israelites to build an ark out of wood and gold, with God supposedly giving very precise instructions."[9]

The plans were precise: "Have them make an ark of acacia wood—two and a half cubits [3.75 feet or 1.1 meters] long, a cubit and a half [2.25 feet or 0.7 meters] wide, and a cubit and a half [2.25 feet] high. Overlay it with pure gold, both inside and out, and make a gold molding around it" (Exodus 25:10–11)."[10] See what I mean? The Ark was incredibly small. As for the Battle of Jericho, and how the Ark played an amazing role, read on.

None of Those Stories Should Be Taken Literally

For our purposes, the most important, and most famous, part of the story of the Battle of Jericho is that the city's huge, stone walls were brought to the ground by trumpets: *sound.* Real trumpets? Almost certainly not! But, definitely *something* of an acoustic nature—and of a dangerous nature, also. Like so many of the legendary stories in this book, time has altered and eroded the original stories, and to the point where papyrus could raise massive stones into the air in Egypt, to where whistling could do likewise at Uxmal, and how mysterious giants—playing music—were able to create Stonehenge. *None* of those stories should be taken literally. *All* of them, though, should be accepted as distorted legends and tales of acoustic levitation. That includes what went down, quite literally, at Jericho. It's important to note that the debate of what the Ark of the Covenant was—and, perhaps, still is—very much survives. Indeed, the investigations are far from being over.

Israelite Priests Were Trained to Manufacture and Use the Lord's Specified Mixture

Then, there's Roger Isaacs. His book, *Talking with God: The Radioactive Ark of the Testimony. Communication Through It. Protection from It*, makes for intriguing reading. He says of his Ark research, and its connection to radiation:

> *Throughout history consumers of incense have used the sweet, smoky fragrance for mystical rites—but not the ancient Israelites. For them incense had a very practical, protective function relative to the Ark of the Testimony. The Israelite priests were trained to manufacture and use the Lord's specified mixture, not to propitiate the gods, not to make a nice smell, not to drive away demons or please kings and pharaohs.*[11]

Isaacs also says: "Incense was used to protect the priests and people from radiation burn. The resinous material had to be burned to become activated. It was the protective smoke, not the fragrance, that made incense effective."[12]

Others suggest that electricity is the key to understanding the mysterious—and very often lethal—properties of the Ark of the Covenant. They just might be right.

Killed by a High-Tension Discharge

Onto the Ark and electricity, David Meyer writes these words: "For centuries, the Ark was viewed as a mystical object from God, beyond the knowledge of man. However, that changed on September 9, 1915, when the famous scientist Nikola Tesla published an article entitled "The Wonder World to be Created by Electricity."[13]

As for Tesla's own article, itself, in part it says: "Moses was undoubtedly a practical and skillful electrician far in advance of his time. The *Bible* describes precisely and minutely arrangements constituting a machine in which electricity was generated by friction of air against silk curtains and stored in a box constructed like a condenser. It is very plausible to assume that the sons of Aaron were killed by a high tension discharge."[14]

There's no doubt that one of the most important aspects of this story—if not *the* most important one—is this: Where is the Ark of the Covenant now? Is it hidden? Could it be lost? Was it destroyed centuries ago?

The Biggest Problem in Solving the Riddle: Today, the Ark Cannot Be Found

Author Mark Pilkington makes it clear that it's not at all implausible that electricity was in use, and in the Middle East, thousands of years ago. He used what is termed the "Baghdad Battery" to make his point:

> *The enigmatic vessel was unearthed by the German archaeologist Wilhelm Koenig in the late 1930s, either in the National Museum or in a grave at Khujut Rabu, a Parthian (224BC–AD226) site near Baghdad (accounts differ). The corroded earthenware jar contained a copper cylinder, which itself encased an iron rod, all sealed with asphalt. Koenig recognized it as a battery and identified several more specimens from fragments found in the region.*[15]

It's tempting, and admittedly totally predictable, to muse on the possibility that just like in *Indiana Jones and the Raiders of the Last Ark*, the Ark of the Covenant might be hidden away somewhere in a government warehouse, never to be opened—chiefly because, today, we don't understand the dangerous technology of earlier, forgotten times. So the best thing to do? Bury it for good. That may be the reason why, today, we only have a very small handle on the science of acoustic levitation. The secrets remain secrets. And, sadly, the incredible truth has been distorted from high-tech antigravity technology to wild tales of magical trumpets that could collapse city walls.

I know which one I go with: acoustics. It's called "science."

CHAPTER 12

WHAT THE RESEARCHERS THOUGHT OF THE ANTIGRAVITY PHENOMENON

It is deeply important to know that matters relative to acoustic levitation, to related ancient mysteries, and to antigravity were all in circulation decades ago. Indeed, much of the research in this particular field was undertaken from the 1950s to the 1970s. To a degree, and after that, talk of antigravity dropped off—for a while. One who went out of their way to champion the theory that antigravity was the answer to our world's massive structures was a dowser and a writer named Francis Hitching. His written work included such titles as 1977's *Earth Magic* and 1979's *The Mysterious World: An Atlas of the Unexplained.* Hitching said that at least "some people" were certain that "the defeat of gravity, was obtainable, and that this was how at least some of the great stones were maneuvered so exactly into position. This is not so unlikely as it first seems."[1] In fact, as we've seen, *unlikely* is not a word that we should be using in the slightest.

Hitching wasn't done, though, by any means. Adding to that, Hitching said that there is a "well-documented ceremony" that still continued "in the village of Shiva-pur near Poona in central India where eleven men link arms and dance around a heavy, sacred boulder of stone, chanting the words *quama ali dervish*, he took things even further." After a few minutes of this ceremony got moving, "they merely touch the stone with their fingertips and it rises, apparently unaided, to shoulder level. Whatever the cause of this, it does not exclusively have to be the villagers who achieve the effect. Many tourists have tried it successfully. The common factors are that it must be exactly eleven people, and they must circle and chant."[2]

Flying Vehicles Activated by the Ley-Line Energies

The very title itself of L. Sprague De Camp's book—*The Ancient Engineers: The Builders of Egypt, Babylon, Greece and India—Who Were They and How Did They Do It?*—shows that, at the very least, he wondered how such incredible efforts were successfully, and spectacularly, achieved. All those years ago De Camp, rather notably, was also keenly interested in the matter of ancient acoustics. De Camp made it very clear to all of his readers that "the wave nature of sound" aided in the development of a theory of "acoustics for use when planning the construction of their theaters."[3] De Camp wasn't done, I should stress. He continued: "The acoustics of the theaters at Epidauros [that was an ancient Greek city] are so good that today tourist guides demonstrate them by striking a match while standing in the orchestra; the scratch is easily heard by tourists on the upper benches. Architects also placed vases of bronze or pottery about a theater, with their open ends pointing toward the orchestra, to act as resonators."[4]

From Health to Gods—and Stone

There is, to be sure, a gaping difference between acoustics that played significant roles in Greek theaters centuries ago and acoustic technology that was designed to raise gigantic stones in Egypt and elsewhere. I should note, however, that this particular issue that De Camp brought up—of acoustics being utilized by the ancients in two ways—is *very* important. Later I address how and why the Levitators were absolutely driven by sound, and in multiple ways, too. We shall see how sound was also used thousands of years ago in relation to health, religion, and even warfare. De Camp, perhaps, could have gone further with his comments and observations on the matter of acoustics. Yet, at least he noted the fact that acoustic and ancient times were intertwined.

There is much more in relation to Greece, however. Athanasios Dritsas, on the Greece Is website, put these words forward: "According to Damon, a music theorist of the 5th c. BC and teacher of Socrates and Pericles, music is powerful because it imitates the movements of the soul. . . . Accordingly, the Greeks of the Classical period attributed the gift of music both to Apollo, for its educational value, and to Dionysus, for its cathartic and therapeutic power."[5]

The History of Music and Art Therapy know of what they speak on this topic:

> *In Ancient Greece, music was believed to have a mathematical relationship with the Cosmos. . . . The ancient Greek philosophers thought that music could serve a therapeutic purpose. Patients in manic states were often instructed to listen to the calming music of the flute, while those suffering from depression were prescribed listening to dulcimer music. . . . Healing shrines in Ancient Greece housed hymn specialists as well as physicians.*[6]

Ley Lines Still on the Board

Then, there was Rene Noorbergen, another one who had a fascination for mysteries of times long gone. His 1977 book, *Secrets of the Lost Races,* is vital reading when it comes to the strange theme of this overall story. He said, forthrightly, that druids, and those who came along before them, clearly understood ley lines and "were able to utilize the linear energies for flight."[7] Noorbergen said something else, too—that was as controversial as it was amazing:

On the day a line became 'animated' by a sunrise directly down a path, the currents were directed so as to charge a body to such a degree that it could be levitated and made to move along the path of a specific level of magnetic intensity. Druidic tradition tells of such heroes as Mog Ruith, Bladaud and the magician, Abris, who possessed flying vehicles activated by the ley-line energies and were able to travel in them as far as Greece.[8]

This is, in essence, antigravity, but activated in a very strange way.

The Stones Were Undoubtedly Moved and Transported in a Special Way Unknown to Us

Rene Noorbergen, who was not only an author, but a war correspondent and a journalist, too, had his very own views on the matter, and on the mystery of Stonehenge, too. Noorbergen had his suspicions that there was a levitation angle to one of the most famous stone circles in the world: Stonehenge. He asked an important question: "Is it possible that the surviving science of the antediluvians included a method of overcoming the law of gravity?"[9]

Clearly, Noorbergen had been snared by the theory of acoustic levitation. Interestingly, one of the things that persuaded Noorbergen that Stonehenge was created by antigravity was the fact that many of the multi-ton stones that comprise Stonehenge traveled from Wales to England, a distance of more than 200 miles—and, on top of that, across a range of mountains. Noorbergen simply could not accept that such an incredible thing could be achieved.

Magnets of the Mysterious Kind

Peter Kolosimo, who we got to know earlier in this book, in relation to the possibility that the Pyramids of Egypt may have been created by using antigravity, had much more to say. Just like so many researchers and writers of the 1950s to the 1970s, he, too, particularly, became entranced by the matter of centuries-old antigravity. In his book, *Timeless Earth*, Kolosimo made that obvious:

> *Pliny the Elder states that the architect Dinocrates, a contemporary of Alexander the Great, constructed the vault of the temple of Arsinoe with "magnetic stones" so that idols could be suspended in mid-air. Rufinus of Aquileia, around A.D. 400, refers to magnetism in describing from his own experience the ascension of a disc representing the sun in the great temple of Serapis, near Alexandria; while Lucian (second century A.D.), who was noted for his skepticism, relates that he saw the image of a Syrian deity being raised into the air by its priests.*[10]

Robert Charroux, in *One Hundred Thousand Years of Man's Unknown History*, made it very clear where he his mindset was going: "South American traditions say that in ancient times all men had the power to fly, and that huge stones could be moved without effort. In Egypt, a true priest was recognized by his ability to rise into the air at will. Jacques Weiss, in *La Synarchie*, assures us that the Egyptians used levitation in building the Pyramids."[11]

Charroux, impressed by what he read, also quoted Weiss, who came right to the heart of it all: "The enormous blocks of stone, weighing as much as six hundred tons, were slightly convex on some of their blocks and form an unshakably solid structure. They must have been transported by levitation and put in place with extreme ease."[12]

"Some Way of Transporting and Placing Blocks that We Do Not Know Of"

Charles Berlitz was known most of all for his *very* controversial research into what has become infamously known as the Bermuda Triangle. I won't drag something like the Bermuda Triangle into the story, however.

I'm not at all impressed by the huge amount of garbage that has been said about it. As a kid, though, I thought the Bermuda Triangle was amazing. Now, I see it as just amazing crap. Such is the way views can change when ages and minds change.

Despite what I just said, I *will* provide a statement from Berlitz, a man who was usually known for his sensationalism and very little else. He said something that stood out and was well worth thinking about:

> *The illogical hugeness of these stones remind one of the stones of the pre-Inca forts at Sacsahuaman, Cuzco, and other sites. The pre-Incas seemed to use whatever what was at hand, without special consideration of the size. In like manner,* unless the original builders of the terrace of Baalbek had some way of transporting and placing blocks that we do not know of, would it not have been easier to cut such cyclopean stones and then place them [emphasis mine]?[13]

Berlitz, I have to admit, did make a good point, one that continues to make sense.

"The Priests of Ancient Babylon Were Able to Raise into the Air Heavy Rocks"

Andrew Tomas, whose books included *We Are Not the First*, *On the Shores of Endless Worlds,* and *Atlantis: From Legend to Discovery*, had a long-time fascination for the mysteries of the world's massive stones—and particularly those stones that didn't stay on the ground. That's right: He, too, was someone absolutely drawn to the tales of huge, floating blocks, and the mysteries concerning levitation. Tomas, who was born in St. Petersburg, Russia, and who became an Australian citizen, had a strong feeling that acoustic levitation was at the heart of all this, but unfortunately he was unable to make a solid case. He said:

> *If we could only insulate things against gravitation, they would become weightless. But so far, this has been a fruitless task. Life would be completely transformed if gravitation were conquered. Cars, trains, ships, planes, and fuel rockets, thus being superfluous, would be displayed in museums. Green grass would grow on roads and highways. Houses would float in the air and men fly like birds. However, these crazy days seem to be far away because, although anti-gravitation research is being conducted by some nations, the mystery of gravity has not been solved.*[14]

Words We Could Have Done Without

In 1975, a book was published titled *Footprints on the Sands of Time.* Its author was L.M. Lewis. Although Lewis certainly supported the ancient astronaut issue in that book, everything was all done the wrong way. I'll share with you why Lewis screwed up to a terrible degree. In the book, Lewis wrote:

> *Are we to believe that Near Eastern and Egyptian man suddenly developed a complete and ordered civilization solely by his own efforts? That nomadic savages—occasionally cannibalistic—had become civilized city dwellers and agricultural magnates. That in the space of a comparatively few years the barbarous wanderers had become patriotic nationalists, living in organized communities amid commendable amenities? They must have been taken in hand by a Master Race. If it were otherwise, would they not have continued to exist on the principle of 'walk about' like the aboriginal natives of Australia.*[15]

Lewis most definitely followed on the "ancient astronauts" template that so many other researchers of that period adhered to. The big mistake, however, was in the words he chose to use—and their implications. For example, utilizing the term *Master Race* over and over again (as Lewis most definitely did) hardly gave him any kudos—or, they shouldn't have. It was likewise with his deeply offensive references to those "barbarous wanderers," and to those "nomadic savages." Unfortunately, similar, such inflammatory terminology can still be found in today's world of ancient extraterrestrials research. You probably know the sources I'm talking about.

No doubt, it was Lewis's controversy-filled words that led *Footprints on the Sands of Time* into obscurity. That is a good thing. Deliberately placing people in an inferior category—and all to justify a theory about extraterrestrials from a period long gone—is outrageous.

"There Are about 2.3 Million Blocks in Total"

Lewis, who developed a deep fascination for the pyramids of Egypt at a young age, synthesized his theories in the pages of *Footprints on the Sands of Time*. Strongly suspecting that levitation was the answer to how the Egyptians erected such huge and heavy stones, he pointed out something important: "The Great Pyramid's mass is gigantic. It has been estimated to have a dimensional capacity of approximately eighty-five million cubic feet. This is scarcely surprising, for its base is said to occupy twelve and a half acres, and each of the four sides is 775 feet long. Its height is correspondingly enormous, for it towers 500 feet into the sky above the surrounding desert."[16]

Lewis added:

> *The construction consists of vast layers of immense granite blocks—2,300,000 of them, weighing at least two and a half tons each, piled in 210 level courses. Because of the decreasing area of each tier, the whole structure rises in a series of huge steps. Its weight is not easy to estimate, though it must be about seven million tons!—and all this stone was transported several miles to the building site.*[17]

Today, close to half a century after Lewis—in outrageous, racist fashion—addressed the subject of the Egyptian pyramids, we have even more precise figures:

> *The Pyramid of Giza features three chambers i.e. queen, king, and the unfinished chamber. It is made up of blocks. There are about 2.3 million blocks in total. It is the tallest structure ever made by man (without machines) in about 3,800 years. It has a height of 756 feet. To determine the weight of the pyramid, we consider the weight of each single building block used. Each of the estimated 2.3 million blocks of stone weighs about 2267.96 Kg (about 2.3 tons) each.*[18]

"If We Take the Story at Face Value, What Kind of Levitation Forces Were Involved?"

Let's head in the direction of one William R. Fix. He, at least, provided a balanced approach to the matter of antigravity and the Pyramids at Giza. He said:

> *according to legend, the priests of Heliopolis and ancient Babylon were able to levitate stones that a thousand men could not have lifted. This would certainly have helped some extraordinary engineering problems. Whether the casing blocks were put into position from the bottom up or the top down, levitation seems a necessary ingredient for the project to have been finished at all, and even with levitation, or some other advanced technique, it was still no easy task to produce the precision and care evident in the building. Of course, the levitation theory has its weakness too—skeptics will want a demonstration.*[19]

There are also the words of Stephen Wagner who, in relation to the issue of levitation in ancient Egypt, said:

> *Was it part of an oral history that was passed down from generation to generation in Egypt? The unusual details of the story raise that possibility. Or was this just a fanciful story concocted by a talented writer who—like many who marvel at the pyramids today—concluded that there must have been some extraordinary magical forces employed to build such a magnificent structure?*[20]

Finally, for now, and back in 1974, we have researcher-author Richard Mooney, the author of *Colony Earth*. Mooney said the following, something that provided a completely plausible theory for why we have tragically lost this amazing science: "There is a tradition that appears in the mythology of the Americas that the priests 'made the stones light,' so that they were moved easily. This connects with the legend of levitation, which may have referred originally to an actual technique or device, *long since forgotten* [italics mine]."[21]

Just like Mooney, I have a strong suspicion that the answer to antigravity is *not* hidden by some theoretical, powerful, New World Order-type organization that is suppressing antigravity from the rest of us. Not at all. Rather, from my perspective, it's the length of time that has caused the truth of this sensational story to be largely forgotten and, sadly, to be relegated to unfortunate myth.

As this particular chapter makes it very clear, the matter of levitation has been around for a long time, and has persuaded far more than a few writers and researchers to conclude that the phenomenon is a genuine one—and one that was particularly active thousands of years ago. How curious, when it's clear that we, as a civilization now, are *still* scrambling around for the answers. And for the lost and forgotten secrets.

CHAPTER 13

THE STRANGE STORY OF THE HUMAN LEVITATORS

One of the most fascinating aspects of this overall story is that it doesn't just revolve around gigantic stones that were said to have been subjected to levitation across the centuries. Reportedly, numerous *people* have claimed the power of levitation, too. While startled, more than a few onlookers watched on, no less. Certainly, the most famous—or, rather, infamous—of all the levitators was an extremely controversial man named Daniel Dunglas "D.D." Home. The *PSI Encyclopedia* states of Home that he was "born in Edinburgh [Scotland], in March 1833. . . . Psychic experiences were not unprecedented in Daniel's family. His mother was frequently subject to clairvoyant (often precognitive) visions, and Daniel himself was apparently somewhat precocious in this respect. One vision, for example, announced the death of his mother and the hour at which she would die."[1]

When People Took to the Skies—Allegedly

While Home was at the height of his paranormal activity, he engaged in numerous séances. He dug deep into the world of the occult, the supernatural, and the paranormal, and even sought to contact the dead;

that's quite a body of work to get one's grips into. There was much more than that, though: Home demonstrated, to anyone willing to see, that he could grip hot coals, without any harm to his hands. Musical instruments would play in his presence and to groups of amazed onlookers. The most incredible part of all this was that Home could levitate himself. Or, rather, that was the big claim that he made.

Of course, and as most people have loudly suggested, the most likely scenario for all of this was that Home was nothing but a highly skilled hoaxer, and also someone who was very good at making people see what he wanted them to see. Indeed, there is no doubt that the 19th century Victorian era was certainly notorious for such suspect, paranormal chicanery. Yet, there *are* reasons why we *should*, perhaps, take a much closer look at Home's life and his claimed astounding abilities. They parallel some of the abilities that the ancient Levitators were able to achieve. For that very reason alone, we should take a good look at Home and his very strange, and alternative, "career."

Home was also said to have been able to levitate nothing less than heavy pianos to the ceilings of rooms and to have had the strange skill of causing objects to lose their weight, which is also particularly intriguing. The exact opposite, too: that he could make small objects incredibly heavy. Home could affect the ground, too, to the extent that he could briefly provoke small earthquakes. Startling? Yes. Controversial? Undeniably. Let's see what else we can learn from this man who had—we are told, at least—cracked the secrets of antigravity. Or, who fooled huge numbers of gullible people.

"On Three Separate Occasions Have I Seen Him Raised Completely from the Floor"

One of those who had the chance to see Home "in action" was Sir William Crookes. He was both a chemist and a physicist, and a member of the Fellowship of the Royal Society (F.R.S.), based in London, England. Crookes said, after having seen Home's amazing activities:

This has occurred in my presence on four occasions in darkness; but I will only mention cases in which deductions of reason were confirmed by the sense of sight. On one occasion I witnessed a chair, with a lady sitting on it, rise several inches from the ground. On another occasion the lady knelt on the chair in such manner that the four feet were visible to us. It then rose about three inches, remained suspended for about ten seconds, and then slowly descended.[2]

Crookes had far more to reveal, too, and more fans to entertain: "The most striking case of levitation which I have witnessed has been with Mr. Home. On three separate occasions have I seen him raised completely from the floor of the room. On each occasion I had full opportunity of watching the occurrence as it was taking place. There are at least a hundred recorded instances of Mr. Home's rising from the ground."[3]

Crookes said in the pages of *The Quarterly Journal of Science* (January 1874), "[T]here are least a hundred recorded instances of Mr. Home's rising from the ground." Crookes added: "On three separate occasions I have seen him raised completely from the floor of the room."[4]

For Crookes, it was all amazing and undeniably incredible. And, undeniably, that was the response that Home was looking for, whether he was a hoaxer or the real thing.

"I Was Simply Levitated and Lowered to My Old Place"

Now, we have the testimony of Lord Lindsay. He, too, was someone who found himself utterly captivated by Home. Lindsay wrote these words of his time spent with Home:

I was sitting with Mr. Home and Lord Adare and a cousin of his. During the sitting Mr. Home went into a trance, and in that state was carried out of the window in the room next to where we were, and was brought in at our window. The distance between the windows was about seven feet six inches, and there was not the slightest foothold between them, nor was there more than a twelve-inch projection to each window, which served as a ledge to put flowers on. We heard the window in the next room lifted up, and almost immediately after we saw Home floating in air outside our window.[5]

Others went on to confirm such similar situations, too. One was a man named William Stainton Moses. He was born in 1839 and became a priest of the Church of England. His books included *Spirit Teachings*, *Spirit Identity* and *Higher Aspects of Spiritualism*. Moses graphically recalled his encounter with Home, which took place in August 1872: "I was carried up when I became stationary. I made a mark [with a lead pencil] on the wall opposite to my chest. From the position of the mark on the wall it is clear that my head must have been close to the ceiling. I was simply levitated and lowered to my old place."[6]

"He Walked on the Lake"

Some of the activities of the leading levitators were performed by the most famous and renowned figures in the world of religion. It wasn't all down to D.D. Home. The Bible says the following:

Right away Jesus made the disciples get into the boat. He had them go on ahead of Him to the other side of the Sea of Galilee. Then He sent the crowd away. After He had sent them away, He went up on a mountainside by Himself to pray. When evening came, He was there alone. The boat was already a long way from land. It was being pounded by the waves because the wind was blowing against it. Early in the morning, Jesus went out to the disciples. He walked on the lake [emphasis mine]. *They saw Him walking on the lake and were terrified. "It's a ghost!" they said. And they cried out in fear. Right away Jesus called out to them, "Be brave! It is I. Don't be afraid." (Matthew 14:22–33)*

Moving on, St. Francis of Assisi was regularly said to have been seen "suspended above the earth, sometimes to a height of three, sometimes to a height of four cubits."[8] Montague Summers, author of the renowned 1926 book *The History of Witchcraft and Demonology*, said that "S. Ignatius Loyola whilst deeply contemplative was seen by John Pascal to be raised more than a foot from the pavement."[7] Summers also said that: "S. Teresa and S. John of the Cross were levitated in concurrent ecstasies in the shady locutorio of the Encarnacion, as was witnessed by Beatriz of Jesus and the whole convent of nuns; S. Alphonsus Liguori whilst preaching in the church of S. John Baptist at Foggia was lifted before the eyes of the whole congregation several feet from the ground."[9]

"Three Feet off the Ground"

Such activity can be found in shamanic rites and rituals, too. The *Psi Encyclopedia* gets to the heart of this angle of levitation: "In shamanic culture, levitation is thought to be involved in the 'journey to the spirit world' The 'first shaman' was said to have been closest to the gods and therefore shared much of their power, including the power to levitate. Shamanistic traditions agree that levitation and other 'paranormal' powers, as they are labeled today, are caused by spirits."[10]

Finally for this particular chapter, and back to the man who—more than anyone else—championed human levitation, D. D. Home, we have to ask: Was he just an outrageous hoaxer? Could he have been someone who conned huge numbers of gullible people? Or could he really have discovered something that was so incredibly ancient and amazingly secret?

In a later chapter we will see how someone else may have done something very similar to what D.D. Home managed to achieve, but decades later, and whose secrets of levitation stayed firmly with him—just like Home. This man was one Edward Leedskalnin. His story is just as amazing, and as secrecy-filled as it is controversial.

CHAPTER 14

A TRIP TO TIBET, NEVER FORGOTTEN

In my 2017 book, *The Slenderman Mysteries*, I wrote the following words:

> *Born in 1868, Alexandra David-Neel had a rich and fulfilling life; it was a life that lasted for just short of an incredible 101 years. She was someone for whom just about every day was filled with adventure and excitement. She was a disciple of Buddhism, was strongly drawn to the concept of anarchy, and had a particular affinity with Tibet and its people, much of which is described in her acclaimed 1929 book,* Magic and Mystery in Tibet. *It's a fine and entertaining tale of road-trip proportions and with a large dose of the supernatural thrown in.*[1]

Tricycle: The Buddhist Review expands on this particular matter:

> *It was not until middle age that she married Philippe Neel, the French engineer who supported her through her subsequent adventures, but with whom she almost never lived. Neel did not understand his wife's interest in Buddhism and the East, but in 1910 he offered her a "long voyage"—he meant something like a*

year—to "get it out of her system." She was gone for fourteen years, traveling and living in India, Sikkim, Nepal, Bhutan, China, Japan, making forays into the forbidden kingdom of Tibet. On her return to Europe, she was celebrated as an adventuress and lived another 45 years as a lecturer and writer.[2]

Alexandra David-Neel: an expert on Tibetan levitation (Wikimedia Commons)

"Her Strange Experiences with Levitation"

There is more, too: While on her adventures and travels, David-Neel came across the world of nothing less than levitation. One of those who was deeply interested in the life of David-Neel was Andrew

Tomas, a man who was *also* intrigued by ancient levitation. Born in St. Petersburg, Russia, he was someone who had a deep interest in the matter of what today has become known popularly (and now, undeniably, tiresome, too) as ancient aliens. His books include *Atlantis: From Legend to Discovery, On the Shores of Endless Worlds, We Are Not the First,* and *Shambhala: Oasis of Light.* Tomas said that David-Neel wrote about her strange experiences with levitation in Tibet, where she lived for fourteen years.[3]

David-Neel did indeed come to realize that levitation was not just a myth. Rather, she came to believe that it was nothing but utter reality.

While on her travels, Alexandra David-Neel made an interesting statement, one that most assuredly ties in with this story: "Setting aside exaggeration, I am convinced from my limited experiences and what I have heard from trustworthy lamas, that one reaches a condition in which one does not feel the weight of one's body."[4]

Expanding on that, Tomas said something important:

> *Actually, the [Belgian-]French explorer was fortunate to see a sleeping lama, or lung-gom-pa. These lamas become almost weightless and glide in the air after a long period of training. The lama she saw in her journey in north Tibet leaped with "the elasticity of a ball and rebounded each time his feet touched the ground." Reading these words, one is reminded of Armstrong's "kangaroo walk" on the Moon!*[5]

The late ancient astronauts researcher Daniel Cohen said of all this: "Tomas believes that the ancients had somehow or another learned to master 'gravitational anomalies' which are not 'uncommon on our planet."[6]

The Reversal of Gravity's Force

Back to Nepal, there is a fascinating story from just after the dawning of the 1950s. It involves a story given to one E.A. Smythies. He was an advisor to the government of Nepal and was told of the story by a servant to that same government. Smythies told the whole story. It went as follows:

> *His head and body were shaking and quivering, his face appeared with sweat, and he was making the most extraordinary noises. He seemed to me obviously unconscious of what he was doing or that a circle of rather frightened servants—and myself, were looking at him through the open door at about eight or ten feet distance. This went on for about ten minutes or a quarter of an hour, when suddenly (with his legs crossed and his hands clasped) he rose about two feet in the air, and after about a second bumped down hard on the floor. This happened again twice, exactly the same except that his hands and legs became separated. The episode was not premeditated and Mr. Smythies was stunned to see this phenomenon of the reversal of gravity's force.*[7]

"Buddhist Mystics Who Have Taken Off in Joyful Flight"

Moving on further, there is the fascinating story of the Oglethorpe University Museum of Art. They note this: "Traditional Tibetan literature similarly tells of Buddhist mystics who have taken off in joyful flight. Buddha himself is said to have done so on several occasions, as did Indian masters such as Nagarjuna and Padma Sambhava. The legacy was adopted by Tibetan mystics in the eighth century, with the yogini Yeshey Tsogyal as a prime example, and continued over the centuries."[8]

We still aren't finished when it comes to Tibet and antigravity. The place, years ago, was filled with tales of incredible antigravity.

"A Whole Assembly of Drumming and Trumpet-Playing Monks Levitate a Large Stone Block"

Writer-researcher Andrew Collins says there is evidence that

> *ultrasonic devices were used to disintegrate rock in Tibetan monasteries, while in the nineteenth century a maverick American scientist named John Ernest Worrell Keely developed sympathetic vibratory apparatus that could raise heavy objects into the air and disintegrate granite. Like the stone cores from Egypt, Keely found that ultrasonics could penetrate quartz much faster than other types of mineral because it so closely matched the ultrasonic frequency range used in this process.*[9]

Ben Joffe, at the Savage Minds website, however, suggests that we should take some degree of caution in relation to all of this. He says:

> *The text from which Collins cites, a book by Swedish airplane designer Henry Kjellson published in 1961 is rich in detail—the account from the Swede Dr. Jarl, who supposedly witnessed a whole assembly of drumming and trumpet-playing monks levitate a large stone block from a meadow to the top of a cliff 250 meters high, is replete with specific numbers and measurements, and is accompanied by various precise diagrams drawn by Kjellson.*[10]

So far, none of the above claims and stories of Tibetan monasteries and of levitated stones have been fully, 100-percent confirmed. Should, then, the data be wholly dismissed? No, not at all. It should be added to the growing body of material—from ancient times to the present day—that is highly suggestive of levitation in centuries ago, even if it has yet to be fully confirmed.

Most notable of all, Collins said that he uncovered data showing that, as late as the 1900s, certain monasteries in Tibet harbored "a sonic technology that included the creation of weightlessness in stone blocks."[11] Those are fascinating words to end this particular chapter on.

CHAPTER 15

FROM "SWEET SIXTEEN" TO THE STRANGEST STONES

How about a giant leap into what we might call relatively recent times? How about an almost magical place that, over the years, has had several different titles, that has had more than a few stone-based secrets attached to it, and that is most famously known as Coral Castle? Has that caught your attention? It surely will, if it hasn't yet. The story is an absolutely fascinating one: a heady mixture of mysteries, of lost love, of doomed romance, and of undeniable intrigue. It all circulates around a man, now long gone from this world, by the name of Edward Leedskalnin.

The story is about his wealth of secrets, too, and of an absolute bevy of giant stones that proved to be just as mysterious as those that were used to create Stonehenge, Baalbek, Easter Island, and elsewhere. Like any strange and complicated, but fascinating, story, we start at the very beginning.

The Mystery Begins

The team behind *Roadside America* say the following of the strange tale and set the strange scene:

Coral Castle doesn't look much like a castle, but that hasn't discouraged generations of tourists from wanting to see it. That's because it was built by one man alone, Ed Leedskalnin, an immigrant who single-handedly and mysteriously excavated, carved and erected over 2.2 million pounds of coral rock to build this place, even though he stood only five feet tall and weighed a mere 100 pounds.[1]

No wonder the story is a totally wild one. It's one of those churning tales that seems to be too good to be true. Upon investigating, we see that it's all too real. And there is a significant chance that the man who got everything going—Leedskalnin, himself—knew precisely how the Egyptians got those huge Pyramids constructed, and with a small amount of effort. Like so many others, though, Leedskalnin decided it wise to keep those secrets . . . well, secret.

From Lost Love to the Secrets of the Stones[2]

A citizen of Latvia, a country on the Baltic Sea and sandwiched between Lithuania and Estonia, Ed had a distinct travel bug. He spent some time in Canada, across much of Europe, and, finally, in the United States, which was where he chose to settle. Before all of that, however, there was something else, something *very* important: the plans that Ed and his beloved girlfriend, Agnes Skuvst, had for their looming marriage. All was going very well. Until, that is, when everything began to collapse. Irreversibly. Whether it was due to Ed being about a decade older than Agnes, or them having very different plans for their futures, we cannot really say for sure. It seems likely, however, that it was a combination of the two, as she was sixteen and he was already halfway through his twenties when they were talking of marriage.

That Ed was hardly wealthy didn't help, at all, either. And Ed never sat still. That wasn't what Agnes wanted. She wanted a normal husband-and-wife life—and for the rest of their lives. Ed, though, was constantly on the road or teaching himself advancing the field of science. As just one example of what could have been many, Ed, in quick time, taught himself nothing less than advanced physics. But Agnes didn't really care about that. He had a fascination for ancient Egypt and for the Giza pyramid complex. Agnes wasn't interested in the slightest. What Agnes liked, Ed didn't. And vice versa. You get the awkward picture?

Just one day before the doomed, ill-fated marriage was set to go ahead, Agnes walked away, vowing never to marry Ed. Maybe she ran, rather than walked; we don't know. No one could blame her, though. Ed was hardly the average guy in his twenties. We know that Ed was crushed, though. By all accounts, however, Agnes was most definitely *not* crushed. She began another life. Ed had no role in it. The whole thing was a tragedy.

The Strange Affair of the Mysteries of Florida's Magnets

Ed, it should be known, was determined to embrace his new life in the United States. Even that was a fraught situation, though—for a while, anyway. It wasn't too long before Ed came down with a very serious case of tuberculosis. Here are when and where the mysteries started to begin. Supposedly, the sheer extent of the tuberculosis was considered by his doctor to be nothing less than fatal. Yet Ed valiantly fought it off with the use of what he said were nothing but magnets. As an aside, you'll remember that David Farrant, while dabbling at the Kit's Coty House in England in 2000, used magnets to make large stones spin around in chaotic fashion. Both men had a thing for magnets. Both men had the ability to move large stones. Back to the story.

When Ed said that he had cleared himself of tuberculosis, no one really knew what that meant; certainly not his doctor, who frowned on fringe and alternative medicine. His words are still not entirely clear now. Whether the magnets worked or not, the fact is that Ed came down with—wait for it—a *second* case of tuberculosis. That was not good news. He fought that bout off, too, also with his mysterious magnets, at least he claimed. When Ed began to get his health back, it was time for a new location and a new life. He chose the inviting, and hot, environment of the small Florida City. Florida is where Ed was destined to spend the rest of his life, and in a very strange, mystery-filled way. A legend was about to begin—a legend that people still flock to learn about.

Writer Steve Winston says that in 1923

> *Leedskalnin began carving from the coral found under the soil of his property. During the next 28 years, he diligently and secretly sculpted 1,100 tons of coral into an open-air "castle," a structure that became a shrine to Agnes. To ensure total privacy, he worked only at night, and filled in the cracks between the stones in the wall—the largest stone weighed 29 tons—with smaller stones so no one could see in during the day.*[3]

"Far from Prying Eyes"

Leedskalnin wasn't finished there, though. In the mid-1930s, he decided to move on, and, as a result, he headed off to new turf. Not too far away, though: This time it was to Homestead, Florida. Though situated barely ten miles from his original location, such was the incredible amount of carved stone that Ed had to work with, it took a whopping three years before all of the stone was finally in its new place. That same place became a definitive tourist attraction, something that was hardly a big surprise. The location was filled with secrets, too.

Random Times stated of Ed:

> *Equipped only with a few lanterns and a few manual objects that he used as a child, when he helped his father with the work of stonemason, in 1923 he started the construction of what was called the Rock Gate Park, now known as Coral Castle: about 1,100 tons of rock were raised by the man to erect those monoliths that would have been part of the stone village he wanted to*

dedicate to his former girlfriend. The man, who never allowed anyone to enter his building site, always worked alone, and far from prying eyes.[4]

"I Have Discovered the Secrets of the Pyramids"

The overwhelming secrecy that Ed was so obsessive and careful about began to make people wonder just how he *was* able to move such massive blocks of stone, and all on his absolute lonesome. For example, he built his very own obelisk that weighed in at an impressive 22 tons.[5] Try moving something like that overnight—and on your own! An enigma was most definitely growing. And growing. And growing. Other huge, carved structures were put into place, too, all by Ed, of course. And, one would imagine that all of them would have required a massive amount of manpower. Well, yes, they *should* have. Somehow, however, Ed didn't need any of that. He did it all alone; *he* was the manpower and nobody else was needed.

The official website of Coral Castle notes that when anyone ever tried to uncover Ed's secrets, he "would only reply that he understood the laws of weight and leverage well."[6] Ed was clearly, then, a man of very few words. There were also these following words from Ed that are highly relevant to the themes of just about this entire book: "*I have discovered the secrets of the pyramids, and have found out how the Egyptians and the ancient builders in Peru, Yucatan, and Asia, with only primitive tools, raised and set in place blocks of stone weighing many tons* [italics mine]."[7]

Culture Trip adds intriguing data to all of this fascinating story:

> *Though he came from a family of stonemasons, [Leedskalnin's] methods have stumped engineers who have compared his techniques to that of the amazing Stonehenge. Some of the stones are even calibrated according to celestial alignments, and the stone walls fit seamlessly together without the use of materials like cement. Leedskalnin once claimed that he discovered not only how*

the ancient builders in Peru and Egypt were able to set in place blocks of stones with only primitive tools but also the secrets of Antigravity.[8]

"Singing to the Stones"

ABCNews offers this: "Leedskalnin was a self-taught expert on magnetic currents, and one theory holds that he positioned the site to be perfectly aligned with Earth's poles to eliminate the forces of gravity, allowing him to move stones weighing several tons each."[9]

Still on this matter, writer Ben Radford says: "Many stories and wild theories emerged over the decades about Leedskalnin and how he built his castle. Some say he levitated the blocks with psychic powers, or by singing to the stones. Others suggest Leedskalnin had arcane knowledge of magnetism and so-called 'earth energies.'"[10]

For all of his amazing—and undeniably mysterious—work, Ed was not destined for a long life. The Reaper was on his way.

The Castle that Never Gave Up Its Secrets[11]

In November 1951 Ed fell sick. As fate would have it, he would not beat this latest bout of illness. It certainly didn't help that instead of calling for an ambulance, Ed decided to take a bus to the Miami-based Jackson Memorial Hospital. Who knows if a quick ambulance drive to the hospital might have saved Ed's life? Possibly. He lingered for a while. Approximately, however, a month after constant illness, Ed was dead from a severe kidney infection. The infection had expanded right up until death coldly took him. Ed was just sixty-four. And he took his secrets of those massive stone blocks with him. Ed's legend, however, most definitely continues on.

To this day, people flock to Coral Castle. They all come back amazed and impressed by Ed's secret work. That's not all: In 1986, rocker Billy

Idol, who has a fascination for Ed's life and story, wrote a song about the man himself, Agnes, and Coral Castle, titled "Sweet Sixteen." And Idol's autobiography, 2014's *Dancing with Myself*, has half-a-page about Ed and his life and career. The story of Ed and his "castle" was told in Season 2 of *Ancient Aliens*. And the location was used for the 1958 cult-classic movie *The Wild Women of Wongo*. Collectively, all of this ensures that Ed's legend will continue to thrive.

Perhaps we should end this very strange, mystery-filled story with the thought-provoking words of *Mysterious Trip*: "The question is still here—how a man 5 feet tall was able to build such structures and moved enormous rocks just like that and make a beautiful coral castle out of it?"[11]

CHAPTER 16

ON THE ROAD AND THE U.S. GOVERNMENT GETS INVOLVED

Moving on from the mysteries of Coral Castle, it is very important to note that although the story this book tells is primarily cemented in the distant past, the fact is that a wealth of research and investigations into the field was undertaken in the 20th century. Time wise, Edward Leedskalnin was a definitive part of that same period: He made it into 1951, but that was it. In this part of the story, begin with one Morris Ketchum Jessup. He was a man who was heavily into the UFO phenomenon (his books were *The UFO Annual, UFOs and the Bible, The Case for the UFO,* and *The Expanding Case for the UFO*). Jessup was also fascinated by the mysteries that surrounded the issue of ancient levitation. Not only that, Jessup's ability allowed him to travel here and there—at least, until his life began to collapse, physically and psychologically, and a series of very strange things caused him, in 1959, to take his life.

Some researchers in the field of conspiracies and flying saucers back then, though, were completely sure that Jessup's death was far more disturbing and sinister. Even to this day, there are still beliefs in the world of cover-ups and secrets that Jessup was deliberately killed for what he knew about the burgeoning secrets of antigravity. Murdered by the sinister, deadly Men in Black? A hired assassin? Who knows? In other words, we are looking at a strange, twisting saga that would have been perfect for *The X-Files* when it was in its prime. With all of that said, let's take a look at Jessup and his intriguing, short life on Earth, and his obsession for levitation and those mysterious people who, in the very ancient past, had the incredible abilities to control it.

From University to UFOs

Born just outside of Rockville, Indiana, Morris Jessup was someone who, as a young boy, gravitated, ahem, toward the growing field of astronomy. Such was his passion for the subject that as time went by, Jessup secured a bachelor of science degree in 1925 at the age of twenty-five, while working at the Lamont-Hussey Observatory, which was operated by the University of Michigan.

Just barely a year later a master of science degree followed for Jessup. Despite all of these impressive achievements, Jessup—somewhat baffling—spent much of his life working as a salesman in parts for cars. That didn't stop Jessup from digging deeply into his other passions, however. One was the field of archaeology, and, when the era of the flying saucer began in 1947, Jessup wasted no time eagerly looking into that issue, too. Of course, the blending of the two issues that Jessup was deeply into—ancient civilizations and UFOs—meant that it was almost inevitable he would become a well-known figure in the UFO scene. Sadly, time was *already* running out for Jessup. He didn't know it, though.

"Not Quite Indiana Jones, but Close"

To his credit, Jessup was certainly not what one would call an "armchair investigator." The man hit the road to a degree that Jack Kerouac and Neal Cassady—combined—would have been proud of. Jessup, though, unfortunately didn't have the kinds of looks that made the girls swoon when Kerouac was around. And he didn't look like Harrison Ford's leather-jacket-wearing Indiana Jones in 1981's *Indiana Jones and the Raiders of the Lost Ark*. Suits, glasses, ties, and jackets were Jessup's things for the somewhat pudgy man who is still spoken about today. But none of that stopped Jessup from doing what he really loved in life: hitting the road and seeking the ancient mysteries of our world. For example, Jessup took road trips to, among other places, Honduras, the Yucatan, Guatemala, Belize, Peru, Chiapas, and Tabasco.

The Great Pyramid of Cholula (also known as Tlachihualtepetl) was one of Jessup's particular favorites, too. There is also another name for this huge pyramid. It is "made-by-hand mountain."

Zaria Gorvett, writing for the BBC, says that Spanish invaders of October 1519

> *arrived in their thousands. Hardened by months of war with ferocious natives, near-starvation and exotic diseases, Hernan Cortez and his Spanish army marched into the great city of Cholula expecting a fight. But this was a sacred city. Instead of investing in weapons, its inhabitants built temples; it was said they had a holy pyramid for every day of the year. After such generosity, their gods would surely protect them. This was a grave error indeed. As the army stormed its streets, religious treasures were looted and the precious pyramids went up in smoke. Within three hours they had murdered 3,000 people.*[1]

"We Find Evidence of Stone Blocks of Unbelievable Weight Being Quarried"

A close and clear reading of Jessup's *The Case for the UFO* makes it very obvious that, by the time his first book was published in 1955, Jessup had all but constructed a complete theory for what lay behind the building of the world's ancient, huge structures. In Jessup's mind, it all just *had* to be down to levitation-driven technology that, by the 1950s, no longer existed on our world. As Jessup saw it, the ancient levitation-based technology was lost or was deliberately hidden thousands of years ago. He wasn't sure which. And it was one of Jessup's roles to try and solve that riddle. One way or another.

Jessup said—while practically giving an unmissable middle-finger to the scientific community:

> *In many areas we find evidence of stone blocks of unbelievable weight being quarried, more or less casually moved considerable distances, then lifted into place. This common factor connects pre-Inca Peru with Easter Island in a startling and undeniable way, and seems to tie in the Middle East, the Orient, Africa, and maybe Polynesia. Many investigators and thinkers have proposed methods for moving these quarried and dressed blocks. All of the proposals are based on application of such simple present-day engineering equipment as block-and-tackle or sand ramps.*[2]

"I Believe the Source of This Lifting or Levitating Power Was Lost Suddenly"

In a 1956 lecture for the National Investigations Committee on Aerial Phenomena (NICAP), Jessup, at the time promoting his book *The Case for the UFO* to a sizeable audience, said:

> *The great pyramids, consisting of hundreds of thousands of huge stone blocks, are thought by some to have been erected by thousands of slaves toiling up long ramps of sand to bring these gigantic masses from the Nile. Flotation has been considered. No suggestions have been made which really fit all cases, and some of the submissions are so cumbersome and inadequate as to seem ridiculous.*

Jessup, speaking to 1950s-era ufologist Gray Barker—the man who just about kicked off the mystery of the Men in Black, by the way—made a concise statement that he used on several occasions, in lectures, primarily, before his 1959 death took him away:

> *I have used the word "levitation" as a substitute for power or force. I have suggested that flying saucers used some means of reacting with the gravitational field. In this way they could apply accelerations or lifting forces to all particles of a body, inside and outside, simultaneously, and not through external force applied by pressure, or harness, to the surface only. I believe that this same, or a similar force was used to move stones in very ancient times. I believe the source of this lifting or levitating power was lost suddenly."*

We come more to the matter of that amazing technology being "lost suddenly" later. That's one of the key issues of this very book: that there is a distinct possibility the science behind the ability to levitate now completely eludes us because no one knows where it is. Lost or hidden, we haven't a clue.

"Paranoia Begins to Take Over for Jessup"

Had Jessup focused his work solely on the matter of UFOs and ancient mysteries, the likelihood is that his research and his book writing would have continued at a steady, probably even merry, pace. Unfortunately, that is not what happened—far from it. That's right: The *exact opposite* took place. Jessup found himself bombarded with letters and late-night phone calls from people who had read both *The Case for the UFO* and, also, 1956's *The Expanding Case for the UFO.* The latter addressed the matter of ancient levitation, too.

I should stress that I get all kinds of weird calls, Facebook messages, and emails, and at all times of the day and night. But, here's the important thing about this: I don't give a shit when some nut-job wakes me up at 3:00 a.m. Yes, I *do* get such things now and again; it's part of the job in this alternative, strange business. I see it all as a part of the UFO game—and I'm totally fine about playing that strange game. I absolutely embrace the craziness that comes my way and that sometimes hits me on a 24/7 basis.

On the other hand, though, for Jessup such similar developments were nothing but stressful and anxiety-creating. It was a situation for Jessup that led him to uncontrollable paranoia. It got even worse when a certain, strange character, Carlos Miguel Allende, came banging on Jessup's front door—not quite literally, but you get what I'm saying. That's when the story of an invisible warship poked its nose in and Jessup was never really okay again. From Uncle Sam's warships to ancient levitation? And a man named Allende? Jessup was beginning to wonder what on Earth he had got himself into. He soon found out. None of it was good. Yes, Jessup really didn't enjoy any of this. It was just *another* layer of pummeling anxiety for the man who was slowly spiraling into a world of insanity.

As for Allende, he was quite a character, to say the very least. Some researchers of this aspect of Jessup's life have suggested that Allende was a definitive whistleblower. For other investigators, though, Allende was just a crazed fantasist; someone who enjoyed conjuring up bizarre stories for his very own amusement. Jessup had a very hard time trying to figure out what was real and what was mere fantasy and nothing else. Jessup didn't stop, however, despite all that stress he simply could not avoid.

"Government Secrets (or, Maybe, *Top* Secrets) Become a Part of the Story"

Taking into consideration the fact that there was no internet in 1955/1956, everything that came to the fore had Jessup wondering what the hell he should do and what was going down. Extremely strange letters sent by Allende to Jessup claimed that the U.S. Navy had managed to successfully, and secretly, make a U.S. Navy ship invisible. The story was that the *USS Eldridge* was the ship made invisible, and that it all went down at the Norfolk, Virginia, naval yard in 1943. So we are told, everything went horrifically wrong: Some of the crew members were said to have been rendered invisible—for the rest of their lives, no less. Others aboard the ship, however, were dying, right before their panicked comrades and friends, on the ship, terrified and, for some, even embedded into the metal of the *Eldridge*. As a result, crew members were hastily, and secretly, "put to sleep."[3] There was no option when a man was buried in the metal. And the terrible story continued to rumble onward.

Jessup was becoming even more worried about what he was getting himself into, particularly when Allende began to expand on the matter of ancient levitation. Jessup wondered: Where was Allende getting all of his material on antigravity? It certainly wasn't from Jessup's own book, *The Case for the UFO*; that much is sure. Allende seemed to know far more than he (Jessup) did about those strange, floating stones of ancient eras.

It was clear to Jessup that Allende was someone who had been deep into the matter of levitation at least a couple of years before Jessup's 1955 book hit the bookstores. It must be said: That didn't mean Allende wasn't a charlatan. It did mean, though, that Allende knew far more than just the basics of levitation—and of how it may have been applied to the building of huge structures all those years ago.

Jessup Gets Deeper and Deeper into an Ever-Growing Puzzle

As for what became famously known as the "Philadelphia Experiment," it was supposedly deemed totally disastrous. As the story goes, the U.S. Navy, unsure of the nature of the technology and the science they had recklessly unleashed, quickly shut everything down, fearful for what might happen. For Jessup, all of this—the claims surrounding the terrible experiment at the naval yard and the words of Allende—might have eventually gone away. And, Jessup, perhaps, could have finally taken some deep breaths and gotten some relaxation. Too bad. That didn't happen. There was a specific reason for that: In 1956, Jessup became a "person of interest" to the U.S. Navy.[4] That's something no one needs to be labeled.

For Jessup, the words *deep* and *shit* very likely swirled around his mind. Panic attacks and over-the-hill anxiety were taking even more toll on Jessup.

The Past and the Present Begin to Fuse Together

The interest that the world of officialdom developed in Morris K. Jessup all began when the U.S. Navy's Office of Naval Research, based in Washington, D.C., received in the mail a copy of Jessup's *The Case for the UFO*. Although, Jessup knew nothing about all that. The book had been posted to the Navy in an anonymous fashion. Whoever mailed the book clearly didn't want the Navy to know who they were. The initial reading of the book went to one Major Darrell Ritter. He could quickly see this was no normal copy of Jessup's book, however. Someone—presumably the person who mailed it to the Navy—had filled it with bizarre scribblings and notes in various colors, making it clear that they knew all about (a) the disaster at the naval yard back in 1943, and (b) ancient levitation. Very intriguingly, the Navy didn't toss the book, or the whole story, out the window. Quite the opposite, in fact. The Navy would soon be consulting Jessup to a significant level.

Behind the scenes, it quickly became clear to the Navy that someone needed to make a deeper investigation of this curious book, of the person who had scrawled those near-endless words in *The Case for the UFO*, and of Jessup himself. Major Ritter wasted no time. He handed over the book, and the entire operation, no less, to a Commander George Hoover and a Captain Sidney Sherby. Notably, they were both in the employ of what was titled the Special Projects Office. That involved the development of fringe-like technology. Meanwhile, at home, Jessup, having been exposed to what seemed to be top-secret Navy experiments, began to worry that one or more agencies of the U.S. government might be following him or that his phone was wiretapped. Maybe both. Whatever, Jessup had reasons to fret. It turned out that Jessup's worries got even bigger.

Dining in D.C. with Naval Intelligence Agents

The time (mid-1950s) was also when the Men in Black began to surface in Ufology: Gray Barker's lengthy 1956 book on the MiB, *They Knew Too Much About Flying Saucers,* was published just one year after Jessup's *The Case for the UFO* hit the shelves. It's no surprise, then, that Jessup had significant worries about the possibility of the dark-suited/fedora-wearing ones knocking on his door late one night. In a near-prophetic, eerie fashion, that's almost what happened. There wasn't a loud banging on Jessup's front door. There was, however, a totally out-of-the-blue phone call from the U.S. Navy: Would Jessup agree to fly out to Washington, D.C., at the cost of the Navy, and discuss the matter of Jessup, Carlos Allende, and his book? They wanted to chat with Jessup for two days, and, in the process, all would be explained to Jessup. At least, he hoped things would be cleared up. Jessup accepted the invite, even though he wasn't sure that he should have or not. Nevertheless, Jessup was soon soaring through the air and wondering what would soon be waiting for him at the other end.

The first thing for Jessup, after landing, was a free dinner, courtesy of the U.S. Navy, in D.C. with Hoover and Sherby. Jessup was relieved there was no good cop/bad cop situation. That's actually what Jessup had been anticipating. It wasn't like that, though: Jessup said later that, surprisingly, everything went down very well. Nevertheless, Jessup couldn't shake off the fact that there was more to all of this. And at times, he even felt that the pair were acting just a tad *too* friendly. No wonder Jessup's mind was spinning like David Farrant's stones.

"Gravity Control, Gravity Nullification, and Levitating Forces"

Levitation and the Philadelphia Experiment were the two issues on the agenda. Not only that, on the following morning, and after a hearty breakfast—once again from the Navy's budget—the two Navy men totally shocked Jessup. They told him the Navy was going to copy no fewer than twenty-five copies of the edition of his book that contained all of those scribblings and notes of Allende about levitation and secret naval experiments run completely awry. There was something else, too: Sherby and Hoover handed over that anonymous copy of Jessup's own book and asked him if he recognized the writing. Jessup was practically floored. It turns out that Jessup *did* recognize the writing: it was the exact writing style of none other than Carlos Allende, Jessup's strange, eccentric informant. The Navy was soon after Allende, who, for many of his years, had either been hiding out or on the road, almost in a hobo style at times.

The job of printing all of those editions for the Navy fell to the Varo Corporation of Garland, Texas. Miss Michael Ann Dunn did all of the work. It wasn't long before those copies were circulated to various people in the Navy who, for whatever reason, were taking so much interest in Jessup's activities, Allende's wild words, and levitation in periods long gone. If the two men from the Navy were concerned that Allende really had hit upon the basics of a top secret U.S. government levitation project, then the words Allende wrote in the Jessup book must have made them outright shudder. They included *gravity control, levitating forces, massive unfinished stones, the secrets of ancient flight, gravity nullification,* and what Allende called the "levitator."

"Encouraging Research in Electrogravitics"

The big irony in all of this is that Allende completely lacked any kind of real credibility; but, even so, he may very well have stumbled upon something all too real—something considered highly classified in the corridors of power. Such a scenario is not at all to be dismissed. Take note of this: In January 1956, writers at *Aero Digest* magazine were provided with a secret, insider tip on how, at that very same time, government agencies in both the UK and the U.S. were working on a highly classified levitation project.5 Unless all of this was a huge coincidence, it seems that Allende *did* know at least *something* about the ways and means to levitate, and he also knew that the U.S. Navy was working on just such a project—a *top-secret* project, no less.

The article, titled "Anti-Gravity Booming," was published in March 1956. In response to the article, one Arthur Valentine "Val" Cleaver, a renowned English rocket engineer, wrote these words: "The Americans have decided to look into the old science-fictional dream of gravity control, or 'Antigravity,' to investigate, both theoretically and (if possible) practically the fundamental nature of gravitational fields and their relationship to electromagnetic and other phenomena."[6]

This is *exactly* some of what Allende was talking to Jessup about more than a year earlier. And it was what the Navy was talking to Jessup about. It's hardly surprising that the Navy was doing its best to figure out the mystery.

"The U.S. Military Tries to Figure Out the Enigma of Antigravity"

Just one year later after Val Cleaver was talking about this, *Product Engineering* said in its pages: "The [U.S.] Air Force is encouraging research in electrogravitics, and many companies and individuals are working on the problem."[7]

Whether or not you or I believe that levitation is a reality—or, that maybe, it *was* in the ancient past—the fact is there's *no* doubt that in the mid-to-late 1950s, both the U.S. Air Force and the U.S. Navy were quietly dabbling in such fringe technology. I say "dabbling" because there's no hard evidence to show that the Navy or the Air Force have ever made significant advances in the field of antigravity. Based on what we've seen so far, it's very much a case of "what happened in the past stayed in the past." That the U.S. Navy quietly chose to enlist Jessup into their ranks, albeit it for just a few days, and were even intrigued, and perhaps even somewhat concerned, about Allende's words—which *also* surfaced in 1956—makes it clear that *something very strange* was going on with the military back in the Fifties. And levitation was right at its heart.

For Jessup, the Final Countdown Wasn't Far Away

Our foray into the strange world of Morris Jessup is almost at its end—for now, at least. It's not a good ending. In fact, it turned out to be an absolutely deadly one. Late in the afternoon of April 29, 1959, at Florida's Dale County Park, the body of Morris Jessup was found, dead, in the driver's seat of his car. The conclusion: suicide. The cause? Carbon-monoxide poisoning. Not everyone, however, was sure that Jessup's death was due to suicide.

Jessup's wife, Rubye, noted that several towels had been pushed into the windows of the car as a means to prevent any carbon monoxide from getting out. So, why didn't Jessup take a few towels from the home to the car? He could easily have done that. Yet, Jessup chose to go to a nearby store and get some new towels. What would be the point in creating such a convoluted situation? On the other hand, though, who can really say what goes through the fraught mind of someone when they are just about to end their life?

There's also the matter of a certain phone call that was made the very night before Jessup's life was snuffed out.

It Was All Over, All Too Soon

As The Unrevealed Files state:

> *A friend of Jessup, Dr. J. Manson Valentine (an oceanographer, archaeologist, and zoologist) said, "Jessup was very upset during*

the last months of his life and was reaching out more than ever to talk with me or someone who can understand his feelings. It was during these last months that Jessup shared his innermost feelings about the Philadelphia Experiment with me." It's very probable that Dr. Valentine was the last person to talk with Jessup. He had spoken to Jessup on April 20, 1959, and had invited him to dinner; Jessup accepted his invitation but never visited.[8]

It's important, too, to note that that whenever and wherever anyone in the field of UFO research and writing dies, there are *always* rumors and whisperings to the effect that there just *had* to have been suspicious circumstances. I have to say that that says much more about the UFO scene—and some of the people in it—than it does about the ways by which their lives *really* ended.

CHAPTER 17

GOVERNMENT AGENTS AND A MAN WITH AN OBSESSION

Bruce Cathie, who passed away in 2013, was someone we crossed paths with earlier. Now, let's dig deep further into his strange, yet intriguing, world of UFOs, of the Levitators, and of antigravity. For many years, Cathie was a captain with the National Airways Corporation of New Zealand. Chiefly, he flew Fokker Friendship and Boeing 737 planes. After a UFO encounter of his very own kind in Mangere, Auckland, in 1952, Cathie decided to try to solve the UFO mystery all on his lonesome. That was a pretty impressive thing to try and do. Some chance of that, but Cathie did at least his best.

What Cathie *did* discover, however, was what he termed a *world magnetic grid system*: an invisible grid that had its origins in ancient times and that allowed those piloting the UFOs to soar across the Earth, almost as a train on the New York City Subway or on the London Underground would. The imagery is decidedly impressive. But what about the reality of Cathie's theories? Well, it wasn't long before Cathie began to focus all of his UFO research on the "grid" angle. In doing so, he uncovered some incredible data.

Journalist and author Mark Pilkington said of Cathie and his work:

> *Collating all the information he could on the subject, Cathie discovered the work of French ufologist Aimé Michel who, in the early 1950s, proposed that UFOs travelled the world by following straight lines between specific waypoints. As a pilot, the theory made sense to Cathie, who began to plot UFO "flight paths" around New Zealand. Before long, he had charted a complex, radial grid system over the entire country. But UFOs were being spotted all over the world.*[1]

"Inexplicable Stone Ruins Have Been Found"

As time progressed, Cathie expanded his work and theories even more. And the more he bought into the grid angle, the more he developed a growing body of followers. He also developed a body of skeptics. Cathie scarcely cared. He went right on without stopping: "After years of work, I discovered that I could formulate a series of harmonic unified equations which indicated that the whole of physical reality was in fact manifested by a complex pattern of interlocking wave forms. I gradually found that the harmonic values could be applied to all branches of scientific research and atomic theory."[2]

There was something even more incredible, too—something even more relevant to the story of antigravity in times past. Cathie quickly came to the conclusion that this grid system amounted to the technology that allowed the ancients to raise massive stones and place them just about anywhere and everywhere on the planet. Cathie said:

> *A number of inexplicable stone ruins have been found as remnants of past cultures—including those of the Mayas, Incase, Aztecs, and Egyptians. In many cases the ruins indicate patterns that were originally geometric in their design. Reconstruction and measurement have proved beyond doubt that in many instances*

the buildings and structures had inherent in them mathematical concepts which had direct connection with light, gravity, and mass.[3]

Cathie and Conversation with the U.S. Defense Intelligence Agency

One of the most fascinating aspects of Cathie's lifelong work in this field brought him to cross paths with the U.S. government. Unlike Morris K. Jessup, though, Cathie wasn't worried about chatting with "the government." Specifically, it was the Defense Intelligence Agency (DIA). In the words of its own staff, the DIA states:

> *The Defense Intelligence Agency provides military intelligence to warfighters, defense policymakers and force planners in the Department of Defense and the Intelligence Community in support of U.S. military planning and operations, and weapon systems acquisition. DIA's diverse workforce is skilled in military history and doctrine, economics, physics, chemistry, world history, political science, bio-sciences, and computer sciences to name a few. DIA officers travel the world, and meet and work closely with professionals from foreign countries.*[4]

An impressive organization, for sure. Now, let's go back to 1965. That was when the DIA and the Foreign Technology Division of the U.S. Air Force came together, and in relation to the work of Bruce Cathie, too. Timothy Good, a ufologist, and the author of the bestselling UFO-themed 1987 book, *Above Top Secret*, said that "Cathie first approached the U.S. Embassy in Washington in the mid-1960s, since which time the DIA kept a file on him. The earliest documented memo is from Colonel John Burnett, Air Attaché, to the Foreign Technology Division at Wright-Patterson [Air Force Base], dated 26 August 1965."[5]

"A Lean, Wiry New Zealander, with an Apparent above Knowledge of Mathematics"

Colonel Burnett recorded these words in his report: "Captain Cathie visited me for about half an hour. I observed this New Zealander to be not only rational but intelligent and convinced that certain UFOs he and others have seen are from outer space—probably Venus. He hesitated in expressing his beliefs in the Venus origin, explaining that it usually tended to convince people that he was a bit of a crackpot."[6] Most Ufologists can relate to that issue of having their characters besmirched.

Timothy Good also had something to say: "By 1967 Colonel Burnett had been replaced by Colonel Lewis Walker, who seems to have been less impressed with Cathie's ideas than his predecessor. But this did not prevent Walker from forwarding Cathie's material to the DIA at the Pentagon."[7] Now, onto what Colonel Walker had to say about all this.

"An Overall Master Plan Exists by an Alien Race"

The military continued:

> *Captain Cathie is still employed as an aircraft F-27 Friendship pilot by National Airways Corporation. His superiors know of his interest and activity in UFOs and his forthcoming book "Harmonic 33." He has been checked for security reasons and no adverse reports are known. He admits that many people consider him some kind of nut but he persists in his theory. On January 1968 [sic] he came to my office and reported that four UFOs had been detected by the Auckland Air Traffic Control radarscope on January 1968 [sic] at 2335 hours local time.*[8]

The DIA continued:

> *Three objects were 15 miles apart in line, with the fourth object in line 30 miles behind the three. Relative speed was extremely high. In addition, two UFOs—disc-shaped—appeared east of Auckland Airport on the same track as first four. Captain Cathie was asked if official reports were submitted on these sightings, and he said no, that Civil Aviation personnel had been warned not to report any more of these observations. Captain Cathie is a lean, wiry New Zealander, with an apparent above-average knowledge of mathematics. He is intensely sincere in his efforts. He is spending an enormous amount of time and effort trying to prove his theory that an overall master plan exists by an alien race—purpose not defined.*[9]

The Dreaded MiB Tailing Cathie?

Now, the story gets somewhat sinister. The Defense Intelligence Agency papers show that Cathie had put in nothing less than a stern complaint with the Americans, to the effect that (a) he was being watched by U.S. agents (and watched closely), and that (b) he wanted those same agents removed—and quickly, too. Cathie said, but failed to explain how he knew, that the Men in Black–type characters were stationed at the *USS Eltanin.*[10] This suggests that at least a portion of this story has gone missing.

As for the ship itself, it happened to be in port in New Zealand when Cathie went on his rant to the U.S. government. As for the *USS Eltanin* here's a brief description:

> *The USNS Eltanin was launched in 1957 as a noncommissioned Navy cargo ship with a special feature that would prove essential to scientists: she was built with an ice-capable, double hull, and officially classified as an Ice-Breaking Cargo Ship. In August of 1962, she was refitted to perform research in the Southern*

Ocean and reclassified an Oceanographic Research Vessel (T-AGOR-8)—in fact, Eltanin was one of the world's first Antarctic research ships.[11]

The DIA outlined the story of this curious surveillance:

Capt. Cathie said that he had been cleared by the NZ government to pursue his research and that he had a letter to this effect signed by the Prime Minister. He stated that the Member of Parliament from his area, Dr. Findley, had interceded for him and obtained government approval for his work. He then asked the DATT (Defense attaché) to "call your agents off. I have official approval to continue my work. I don't want them tailing me." The DATT made no reply to this request. This man is obsessed with his theory and no amount of argument can convince him that he has not stumbled on a highly complicated system which he says leads directly to the existence of UFOs.[12]

Perhaps Cathie really did discover such a "complicated system." Or maybe there was a fair degree of paranoia, after all, of the type that infected Morris Jessup.

The Final Words on Bruce Cathie

Was Bruce Cathie really secretly watched as a result of his research into such issues as antigravity, levitation, and the Pyramids of Egypt? Or, in a decidedly ironic state, had he become a victim of his very own brand of paranoia? If so, was it paranoia that led him to *believe* he was under surveillance, when the reality of the situation just might have been far different? Who were the "three persons" who "accosted" Cathie in the Grand Hotel in 1968? If he was being watched, then what was the purpose? And what was it that led Cathie to be "checked" for "security reasons"? Taking in consideration the detailed words Cathie had to say about all of this, it is possible there were (and perhaps still are) additional files on this curious aspect of Cathie's life.

Also, we need to remember that it was Cathie himself who approached the U.S. military, rather than the other way around. By that, I mean that the Defense Intelligence Agency was not begging to know the results of Cathie's work and research in antigravity. The DIA personnel assigned to deal with Cathie spent enough time to make it clear they found his theories at least interesting. The DIA did not, though, take enough interest to warrant funding, which may very well have been what Cathie was hoping for. But shouting loud that he had been accosted by U.S. agents would hardly have got Cathie in the DIA's good books.

So certainly, yes, the DIA did follow the work of Cathie, and over a period of several years. There's no hard data, though, that anything sensational came to the fore. Unless, of course, those Men in Black types knew much more. And, based on what we've seen, that might just have been the case.

In this curious caper, the questions are many. Of course, *none* of this proved to the DIA—or, in fact, to any agency of the U.S. government—that the Pyramids of Egypt *were* constructed in fantastic, high-tech fashion. So far, this demonstrates two things and two things only: (a) Bruce Cathie contacted the authorities in relation to his research, and (b) as a result, those same authorities took at least a mild interest in what he had to say. Whether or not more was going on behind the scenes still very much remains to be seen. After all, although the approach the DIA took to Cathie was not particularly exciting, things were *very* different when it came to the U.S. Navy and Morris Jessup. There was clear concern on the part of the Navy as to what Jessup—and possibly Carlos Allende—had found. Or stumbled upon.

CHAPTER 18

AN AERIAL BATTLE OVER ISLANDS

Another 20th-century figure who took a decisive role in the controversy surrounding levitation, in early periods and up until the time of his death in 2009, was John Michell, author of the acclaimed and groundbreaking 1969 book *The View over Atlantis*. Michell, who was tutored at Eton College in Berkshire, England, and who served in the UK's Royal Navy, was someone who had a huge fascination for the ancient history, folklore, and legend of the British Isles. He had a particular interest, too, in what are known now as ley lines.

Elizabeth Borneman, writing for *Geography Realm*, says:

> *The theory of geography that surrounds the existence of ley lines was created in 1921 by a man named Alfred Watkins. . . . While in Herefordshire he stopped to examine a Roman camp that was under excavation and, while looking at the areas surrounding the site, noticed that other ancient monuments such as churches, monoliths, forts, and other artifacts seemed to be placed in relation to one another that suggested a higher level of planning rather than haphazard design.*[1]

"Without Warning it All Happened Suddenly"

The more Watkins looked into the locations of some of the UK's most well-known and legendary sites (and their lesser-known ones, too, it must be made clear), he came to the incredible belief that the landscape of the UK was dotted with ley lines. Possibly everywhere. Watkins also concluded that ancient castles and churches, situated very close to ley lines, played particular, key roles in the navigation of the country by, in essence, a kind of a supernatural map.

Such was the nature of the intriguing alignments, Watkins's research suggested that there was far more to the layout of sacred sites dotted all across the British Isles than just haphazard placements. Allen Watkins, Alfred's son, said of his father's thought-provoking discoveries: "[W]ithout warning it all happened suddenly. His mind was flooded with a rush of images forming one coherent plan. The scales fell from his eyes and he saw that over many long years of prehistory, all trackways were on straight lines marked out by experts on a sighting system. The whole plan of the Old Straight Track stood suddenly revealed."[2]

Ley lines were not just ways and means to navigate the landscape. Certainly, things went further than that. Writer Katie Serena said of all this:

> *To those who do believe in ley lines, the concept is quite simple. Ley lines are lines that crisscross around the globe, like latitudinal and longitudinal lines, that are dotted with monuments and natural landforms, and carry along with them rivers of supernatural energy. Along these lines, at the places they intersect, there are pockets of concentrated energy, that can be harnessed by certain individuals.*[3]

It was this angle of "supernatural energy" that led John Michell to believe this somehow was tied in with the controversy surrounding antigravity, particularly ancient antigravity.

"His Craft Was of Stone"

In a 1973 lecture at Glastonbury, England, where, it has been suggested the remains of none other than the legendary King Arthur were buried, John Michell told his audience the following:

> *There are several legends about the Druids' power of flight. Abaris, a British magician, traveled through the air to Greece with the aid of instrument described as a golden arrow. Bladud, the Druid father of King Lear, crashed his airborne vessel on Ludgate Hill in London, the site of St. Paul Cathedral. Simon Magnus, whose own exhibition of flight was disastrously cut short by St. Peter during a magical contest in Rome, had a Druid pupil called Mog Ruith, who possessed a flying machine and is said to have engaged in an aerial battle over Ireland with a rival Druid. Mog Ruith's flying career, like those of his master and King Bladud, ended in catastrophe.*

Of great significance are the following words of Michell, in relation to Ruith: "It is interesting to hear that *his craft was of ston*e [italics mine]; a portion of it can still be seen in Ireland, for his daughter Tlachtga had it erected as a standing stone at Cnamchoill."

Yet again, we see stone being the key medium to allow for one to take to the air. Soaring through it, even.

Michell Would Go No Further

It's important to note the next part of the story. By making such a head-turning statement, it brought Michell even further to the forefront of this field of levitation in the Seventies. His words read:

> *A great scientific instrument lies sprawled over the entire surface of the globe. At some period – perhaps it was 4,000 years ago—*

almost every corner of the world was visited by a group of men who came with a particular task to accomplish. With the help of some remarkable power, by which they could cut and raise enormous blocks of stone, these men erected vast astronomical instruments, circles of erect pillars, pyramids, underground tunnels, cyclopean alignments, whose course from horizon to horizon was marked by stones, mounds, and earthworks.

There's another angle to all of this, too: In the same way that Morris K. Jessup largely held back from claiming that the Levitators were extraterrestrials, Michell—despite having a deep interest in the UFO enigma—would go no further than call the Levitators "a group of men."

CHAPTER 19

FROM THEN TO NOW—AND THEN TO STONEHENGE AGAIN

One of the strangest events that ties in with acoustic levitation, antigravity, and Stonehenge took place not in the distant past, but as late as 1994. It's a very strange saga that adds more and more weight to the theory that government agencies, to some degree, are carefully and quietly studying the field of antigravity, and are also looking for ways to apply it on the battlefield—if such a thing might one day be needed. If that sounds overly sensational, it's not.

Although the incident I'm talking about took place in 1994, it wasn't exposed by the media until March 1997; such was the level of secrecy surrounding it. It was primarily thanks to the *Independent* newspaper that the story came cascading out. Their headline was undeniably an eye-catcher. And there was no way to shut the doors. The *Independent's* headline was "Secret U.S. Spyplane Crash May Be Kept under Wraps." As for the story itself, well, it was a fascinating one.

As the *Independent* said in its article:

A top-secret United States spyplane which flies on the edge of space at five times the speed of sound crashed at the British experimental airbase at Boscombe Down, Hampshire, in September 1994, according to a report in a leading military aviation journal. The SAS [Special Air Service], the report said, was scrambled to throw a cordon round the wreckage, which was flown back to the US two days later. The hypersonic reconnaissance aircraft, called Astra or Aurora, is believed to have been developed in the 1980s as a secret US government "black program."[1]

The author hanging out at Stonehenge (Nick Redfern)

Trying, but Failing, to Hide the Truth

The United Kingdom's Ministry of Defence (MoD), along with a handful of other government agencies, scrambled to try to snuff out the story. Too bad. They failed—spectacularly. One of the departments that helped to try to hide the story was the National Archives (an arm of the UK government), situated not at all far from the city of London. Its personnel released the following to the nation's media:

> *[T]he only flying that took place that night was the launch of two Royal Navy Sea King helicopters in support of an exercise. Claims that members of the public were turned away by police roadblocks may have arisen from some confusion over dates. On August 12, 1994 a Tornado participating in a trial made an emergency landing there after the decoy target under trial failed to jettison. The Tornado landed with a trailing 375ft steel cable and, for safety reasons, roads close to Boscombe Down were closed while the aircraft passed overhead. We are aware of press reports regarding an aircraft known as "Aurora." The Ministry of Defense has no knowledge of any U.S. aircraft with this designation operating in UK airspace. The existence of such a program would, in any case, be a matter for the US Government to confirm.*

Uh-huh. There were many who went on to disagree with that.

Focusing on the Base

The disinformation program—designed to squash the truth of the incident—failed. Badly. Insiders in both UK and U.S. defense intelligence agencies were extremely careful to make sure the story didn't get out, in an Edward Snowden fashion, one might say. What *really* happened was that roughly an hour before midnight on September 26, 1994, a small, very strange-looking aircraft—that was broadly of a triangular shape—had to make an emergency landing at the base, the aforementioned Boscombe Down. There's no doubt that the base would have been the perfect place for secretly bringing the plane to the ground, and with a minimum of fuss.

Wikiwand says of the facility, itself: "The base was originally conceived, constructed, and operated as Royal Air Force Boscombe Down, more commonly known as RAF Boscombe Down, and since 1939, has evaluated aircraft for use by the British Armed Forces. The airfield has two runways, one 3,212 meters (10,538 ft.) in length, and the second 1,914 meters (6,280 ft.)."[2]

The Story Continued to Circulate, to the Concern of the U.S. and UK Governments

Although a great deal of effort, apparently, went into hiding the story, it wasn't just the press who got onto it. It was the public, too. Unfortunately for the governments of both the United Kingdom and the United States, on the night of the event (and on the following morning, too), a number of aircraft enthusiasts, armed with their binoculars and night-vision equipment, were carefully, and secretly, keeping watch regarding what was afoot. Since those same enthusiasts were just members of the public, they didn't have any access to the base. Military police made that clear to anyone who tried to get close to Boscombe Down when the whole thing erupted. Indeed, both military police and regular police were brought in to ensure that nothing could leak out. And roads were blocked off, no less.

That didn't stop those same air enthusiasts from trying to solve the mystery. In trying to do so, several such enthusiasts were taken into custody and warned not to discuss the affair, and what they knew and saw. It was one of those "or else" situations. Luckily for the government, those guys with their binoculars and night scopes really didn't have much to tell, after all, and didn't see much, either. They knew that *something* had happened, but what it was, and still is, is anyone's guess—apart from those on the inside in government, of course.

The Real Picture Starts to Become Clear

The staff of *Air Forces Monthly* magazine knew a good story when they saw it. And, as a result, they did their very best to try and solve the entire incident. The November 1994 issue of the magazine had a full-length story about the affair at Boscombe Down. One of the rumors, the staff of the magazine said, was that the aircraft was a TR-3A, a plane said not to have even existed. A great deal has been said about this particular aircraft. Some airplane experts have suggested that the aircraft was a next-generation Stealth-type aircraft.

A far more exotic theory, however, suggested that the craft, also said to have been known as the *Black Manta*, was powered by nothing less than our old pal antigravity. Yes, that phenomenon really *does* have a history that spans thousands of years. Then and now. On top of that, there is some merit to the theory that there was an antigravity component to this curious story. Late one night, and no less than just one month before the incident at Boscombe Down occurred, a number of people in the area of the base reported seeing "UFOs" and "flying saucers" hovering in silence—and at perilously low levels—over the A303 road, that just happens to run past next to none other than Stonehenge.

The Media Gets Involved Again

The *Salisbury Times* newspaper wasted no time, at all, in sharing the sensational story with its readers. The article read as follows and in its entirety:

> *A green flying saucer hovered beside the A303 road at Deptford last week—according to a lorry driver who rushed to Salisbury police station in the early hours of the morning. The man banged on the station door in Wilton Road at 1:30 a.m. on Thursday after spotting the saucer suspended in mid-air. "He was 100 per cent convinced it was a UFO," said Inspector Andy Shearing. The man said it was bright green and shaped like a triangle with*

rounded corners. It also had green and white flashing lights. Other drivers had seen it and were flashing their car lights at him. A patrol car took the driver back to the spot but there was no trace of the flying saucer. Inspector Shearing said police had been alerted about similar sightings in the same area in the past.

So deep is this particular story buried, we have very little more to work on. There are, however, a few fascinating points that are worth addressing. For a few years, the story went into a limbo. It didn't stay like that, however, I'm pleased to say.

From Past to Present

In April 2021, The Drive site resurrected this strange story, saying this: "Evaluating the Boscombe Down incident nearly 30 years later is clearly a difficult proposition. Whatever the case is, the evidence available about the incident suggests that *something* sensitive occurred at Boscombe Down on that fall evening back in 1994, but the origin of the incident remains a mystery to this day."

As we wrap up this section of the modern era of antigravity, something important should be noted: as I said, the aforementioned A303 road runs right past Stonehenge. On top of that, the distance from Stonehenge to Boscombe Down is less than seven miles. And Deptford, the site of a silent, hovering "UFO," which may have had some kind of "antigravity drive" at its heart, is just 12 miles from Boscombe Down.

A Whistleblower or a Joker? Maybe a Bit of Both?

The final words (so far, anyway) concern a man named Ralph Noyes. He was someone who, for many years, worked in the UK Ministry of Defence and, while doing so, spent a number of years investigating UFOs, both in and out of the MoD. Whether he was just jesting or not, Noyes said—after the story of the Boscombe incident got out—that the UFO had been "*refueling its engines from above Stonehenge.*"[3] Did Noyes have insider secrets concerning ancient antigravity technology and Stonehenge? Did the the Ministry of Defense have such technology, too, too?

Or, was all of this just a joke on the part of Noyes? Certainly, there is no reason why it could have been a prank. Maybe, though, by making a lighthearted statement, Noyes was trying to tell a sensational, secret truth in a way that would have prevented him from getting into deep shit with his previous employers at the Ministry of Defence? Interestingly (in fact, *very* interestingly) Noyes actually did something *very* similar back in 1986. That year, Noyes wrote a full-length novel titled *A Secret Property*. It was a thinly veiled story of the famous "Rendlesham Forest UFO landing" of December 1980 in England. More than a few UFO researchers (including me) believe that by presenting what he knew in the form of a novel, Noyes could tell the truth about Rendlesham without getting into hot water with his irate superiors in the MoD. Just maybe, Noyes did the very same thing with those eye-opening words *refueling its engines from above Stonehenge.*

Now, how about a look at the current state of acoustic levitation?

CHAPTER 20

ACOUSTIC LEVITATION TODAY

We've addressed the issue of levitation in ancient times, during the Middle Ages, in Victorian times, and even when the Cold War was still afoot. There's just one time frame for us to focus on now: the present day. How do scientists see levitation today, particularly acoustic levitation? That's a good question, considering that although acoustic levitation is a real phenomenon, it has become connected to the "ancient astronauts" controversy. Without revealing their names, I know of two scientists who flatly refuse to bring the "Egypt and antigravity" angle of all this to the table, for fear of having their reputations smeared. Or, even worse, completely destroyed. In some respects, we can't blame them for having concerns. *Big* concerns. Despite my words, the majority of scientists in this arena look at the "ancient astronauts"/acoustic levitation subject with an interesting mind, but largely ignore the kinds of work that people like me are doing. So it goes.

Regardless of what the scientific community thinks of ancient mysteries and stones in the skies, the fact is that, right now, a great deal of work is being done in this particularly controversial phenomenon, even though, in many ways, it baffles us today, but apparently didn't do so, to any degree, in centuries past. Brett Tingley is someone who has taken keen note of acoustic levitation and realizes the important and planet-changing potentials it could have for us, as a civilization. He says of all this: "Acoustic levitation – using sound waves to levitate objects—has been the subject of past experiments and has shown promise in achieving true levitation. However, scientists have previously been limited to levitating very small objects or have only levitated objects along one axis, meaning objects had to be semi-restrained."[1]

And, largely, that's the problem. Right now, even I have to concede that we are still caught in a situation that certainly allows for acoustic levitation to exist, but that completely lacks the much-needed muscle to utilize it in the fashion that the Egyptians, the people of Easter Island, and the creators of Stonehenge did all those years ago.

Progression in the 21st Century

Some of the most important data in this arena can be found in an important document of June 2016. It was prepared by Marco A.B. Andrade, Anne L. Bernassau, and Julio C. Adamowski. Its title: "Acoustic Levitation of a Large Solid Sphere." The words of the groundbreaking trio state the following: "We demonstrate that acoustic levitation can levitate spherical objects much larger than the acoustic wavelength in air."[2]

A Whole Host of Other Wacky Applications

The team at *Mysterious Universe*—for whom I work on a regular basis—asks a question that anyone who has an interest in this field wants answered: "Is levitation real?" The decisive response (hypothetical, to a degree): "Sure it is. Scientists have been levitating objects in laboratory conditions for decades. Magnetic and electrostatic levitation have long been used to create conditions for containerless storage and analysis, for materials that react with other common elements. NASA uses them regularly. But there's another method you might not be aware of. Acoustic levitation is quickly becoming a superstar of fringe science. It is used in lab settings for the same reasons as above, and in the research of sound waves and their effect on the environment. But some say it's been used outside the lab for centuries, and maybe even millennia." Indeed, as we've seen.

There is yet a further advance in the field of levitation; this one came along in 2019. Writer Sequoyah Kennedy said: "[N]ew research from the California Institute of Technology has apparently found a way to levitate macro-scale objects using nothing but light. Scientists at Caltech say that, once implemented, this technology would allow a spacecraft to surf its way on a beam of light to the nearest planet outside our solar system in as little as 20 years."[3]

Undeniably impressive.

A Method that Could Levitate Macroscopic Objects

In an article with the title of "Self-Stabilizing Photonic Levitation and Propulsion of Nanostructured Macroscopic Objects," and penned by Ognjen Ilic and Harry A. Atwater, we learn something very important. Notably, Ilic says: "One can levitate a ping pong ball using a steady stream of air from a hair dryer. But it wouldn't work if the ping pong ball were too big, or if it were too far away from the hair dryer, and so on."[4]

What this demonstrates, though, is that great leaps and bounds are being made in this field. Harry Atwater, a Howard Hughes Professor of Applied Physics and Materials Science at Caltech, says something incredibly important, too. It also gives us food for thought in relation to where, precisely, all of this will be heading in the near future: "We have come up with a method that could levitate macroscopic objects. There is an audaciously interesting application to use this technique as a means for propulsion of a new generation of spacecraft. We're a long way from actually doing that, but we are in the process of testing out the principles."[5]

A Virtual Vortex

Paul Seaburn, of Mysterious Universe, said in 2018: "In a new study published this week in *Physical Review Letters*, lead author Dr. Asier Marzo of the university's department of mechanical engineering reveals how researchers have broken the size barrier that restricted previous levitation devices and tractor beams from raising anything more than tiny particles. Even more impressive, they've managed to control the deafening sound as well."[6]

An Important Step Forward

The particular paper that Seaburn was talking about goes by the title of "Acoustic Virtual Vortices with Tunable Orbital Angular Momentum for Trapping of Mie Particles." In summary, there are these words from the team of Asier Marzo, Mihai Caleap, and Bruce W. Drinkwater:

> *Acoustic vortices can transfer angular momentum and trap particles. Here, we show that particles trapped in airborne acoustic vortices orbit at high speeds, leading to dynamic instability and ejection. We demonstrate stable trapping inside acoustic vortices by generating sequences of short-pulsed vortices of equal helicity but opposite chirality. This produces a "virtual vortex" with an orbital angular momentum that can be tuned independently of the trapping force. We use this method to adjust the rotational speed of particles inside a vortex beam and, for the first time, create three-dimensional acoustics traps for particles of wavelength order.*[7]

Ancient Origins, who are far more than familiar with the mysteries of the distant past, have been keeping a careful look at how all of this is playing out in the 21st century:

In two experiments, scientists have successfully levitated lightweight polystyrene balls greater in size than the wavelengths used to elevate them, which represents an important step forward in the management of the force of concentrated sound. One of these experiments, carried out by a joint team of researchers in the UK and Brazil in 2016, lifted a 50-millimeter polystyrene ball several centimeters off the ground, where it remained suspended for as long as the sound waves were generated.[8]

The U.S. Government and its Interest in Levitation

Now it's time to see what the government of the United States thinks of all this. After all, we know that Morris Jessup was approached by the U.S. Navy in the 1950s and Bruce Cathie had correspondence with the Defense Intelligence Agency. It makes sense that research in this area should still be going on right now. Indeed, it is. To learn what the government knows of acoustic levitation, we have to go knocking on the door of the Department of Energy Office of Scientific and Technical Information. They have found something that's undeniably important:

We demonstrate that acoustic levitation can levitate spherical objects much larger than the acoustic wavelength in air. The acoustic levitation of an expanded polystyrene sphere of 50 mm in diameter, corresponding to 3.6 times the wavelength, is achieved by using three 25 kHz ultrasonic transducers arranged in a tripod fashion. In this configuration, a standing wave is created between the transducers and the sphere.[9]

Also from the government:

The axial acoustic radiation force generated by each transducer on the sphere was modeled numerically as a function of the distance between the sphere and the transducer. The theoretical acoustic radiation force was verified experimentally in a setup consisting of an electronic scale and an ultrasonic transducer mounted on a motorized linear stage. The comparison between the numerical and experimental acoustic radiation forces presents a good agreement.[10]

We still aren't finished with officialdom.

Government Work Continues

The U.S. government's United States National Library of Medicine, a branch of the National Institutes of Health, offers this: "Levitation and manipulation of objects by sound waves have a wide range of applications in chemistry, biology, material sciences, and engineering. However, the current acoustic levitation techniques are mainly restricted to particles that are much smaller than the acoustic wavelength."[11]

Government scientists explain:

In this work, it is shown that acoustic standing waves can be employed to stably levitate an object much larger than the acoustic wavelength in air. The levitation of a large slightly curved object weighting 2.3 g is demonstrated by using a device formed by two 25 kHz ultrasonic Langevin transducers connected to an aluminum plate. The sound wave emitted by the device provides a vertical acoustic radiation force to counteract gravity and a lateral restoring force that ensure horizontal stability to the levitated object. In order to understand the levitation stability, a numerical model based on the finite element method is used to determine the acoustic radiation force that acts on the object.[12]

We Have Total Control of the Acoustic Field Inside

Science Daily, in 2017, got into the matter of levitation, and in a fascinating fashion, no less:

> *Levitation techniques are no longer confined to the laboratory thanks to University of Bristol engineers who have developed an easier way for suspending matter in mid-air by developing a 3D-printed acoustic levitator. This new technique, published in* Review of Scientific Instruments, *could be applied to a range of applications, including blood tests.*
>
> *Anyone who has felt their chest vibrating with the energy of the soundwaves at a festival is already familiar with the principle behind acoustic levitation. Acoustic levitation uses powerful acoustic waves to push particles from all directions and trap them in mid-air. By using ultrasound—a high-pitched sound above human hearing—it is possible to use powerful vibrations without causing any harm to humans.*[13]

This Requires Extremely Powerful Sound to Levitate

Eager to get involved, too, the BBC shared these words with its readers and listeners:

> *Scientists at the Swiss Federal Institute of Technology (ETH) in Zurich, Switzerland have designed an acoustic levitator capable of controlling and mixing substances as they hover between two platforms. The device is built of two facing platforms made up of sound-emitting squares that trap the substances between them. The sound waves move upward until they reach the surface lying above,*

whereupon they bounce back. When upward and downward-moving waves overlap, they cancel out and trap materials in place.

In the next chapter, we'll learn about another form of acoustics that is integral to the overall story: the work of the Levitators. This particular form of acoustics was designed to guard those very same stone circles and ancient structures. It did so in a deeply alternative fashion.

CHAPTER 21

HEADS OF HORROR AND STONE

Here and there in this book I have suggested there is evidence that shows the Levitators didn't work with just antigravity only, but that they also used various *other* types of sound, whether in prayer, in warfare, and relative to health. Let's look at this other side of sound, which still has direct connections to ancient stone.

Get ready for one of the most intriguing, and fear-filled, stories of the early 1970s. It sounds like one of those 1960s/1970s Hammer Productions movies, with actors Peter Cushing and Christopher Lee at the helm. This story is not born out of entertainment fiction, though. You would, however, easily be sucked into thinking that was exactly what it was. It's a story that has mysterious threads and links to the story of nothing less than the ancient Rollright Stones that we have addressed on several occasions earlier, and that is why I have included it in this book. If you thought we were finished with that particular stone circle, you're wrong. There are strange things to come. Soon, they will become terrifying. To fully appreciate the whole, weird saga, we have to go back in time to the dawning of the 20th century.

An example of an old, carved head of stone (Nick Redfern)

Our story begins on December 10, 1904, but, as you'll soon see, it was resurrected—in a very strange way—in 1972. As for that 1904 story, it was splashed across the pages of the *Hexham Courant,* a newspaper situated in the north of England. The eye-catching headline was "Wolf at Large in Allendale." The story went as follows:

> *Local farmers from the village of Allendale, very near to Hexham, had reported the loss of their livestock, so serious that many sheep were being stabled at night to protect them. A shepherd found two of his flock slaughtered, one with its entrails hanging out, and all that remained of the other was its head and horns. Many of the sheep had been bitten about the neck and the legs—common with an attack made by a wolf.*

You might well wonder what on Earth a story on antigravity, sound, sonics, and levitation has a place within these pages. We're nearly there.

Something on the Loose

The newspaper story continued and, as a result, had just about all of the locals in states of downright fear, as the story makes very clear:

> *Hysteria soon set in. During the night, lanterns were kept burning to scare away the wolf, and women and children were ordered to keep to the busy roads and be home before dusk. The "Hexham Wolf Committee" was soon set up to organize search parties and hunts to bring down the beast using specialized hunting dogs, the "Haydon Hounds," but even they could not find the wolf. The Wolf Committee took the next step and hired Mr. W. Briddick, a trained tracker. But he was also unsuccessful, despite searching the woods.*

Just a couple of weeks after Christmas 1905, the story grew even more. And, it spread across the countryside. The corpse of a full-grown wolf was found on a rail-track at the village of Cumwinton, a tiny locale about 30 miles from where the initial savage activity occurred. Things got even weirder when it was discovered that this was a second wolf, rather than the first one roaming around the area and killing sheep. The story grew, to the extent that the locals began to wonder if the area was now infested with an entire pack of wolves. Such a thing would have been amazing for one key reason: the last wolf in the UK was killed in the Scottish Highlands way back in 1680. There were other rumors, too. One of those rumors was that the wolves were of a supernatural nature; creatures that had the ability to take on different forms and even dematerialize. And, in a very odd way, they had a connection to ancient stone.

And in an even weirder way, the tale was resurrected in the 1970s.

Strange Heads Found under the Ground

The story of the Hexham wolves (or rather, of the *paranormal* wolves of Hexham) faded away, as such stories so often do. In a fashion, however, this story returned—and in a spectacularly horrific fashion. It got its claws into just about anything that it could—almost literally, at least, to a degree in the early part of 1972. This was when the events of 1904–1905 blended in with strange activity in the long-gone Seventies. It was also when a story of strange stones appeared.

It all began with two young Hexham boys (neither of them had even reached their teens at the time): Colin Robson and Leslie Robson. On one day in February 1972, the two lads decided to play around and dig up a bit of the backyard, as kids tend to do. In doing so they stumbled upon something undeniably amazing. It was something wholly unanticipated. It was a pair of what looked like very ancient, carved heads of stone, both roughly about the size of a tennis ball. One looked female, and the other had a male appearance.[1] The boys' excitement levels practically went through the roof, as they would for a pair of boys looking for fun. Interestingly, the female-looking stone head had a creepy crone-style appearance to it. The story became even more exciting when the boys saw that it even had a hooked, witchy nose. The other freaky face, archaeologists and historians suggested when they were finally able to get their hands on the heads, had the hairstyle of the Celts, who were very much long gone.

The *World History Encyclopedia* provides us the following on these long-gone people: "The ancient Celts were various tribal groups living in parts of western and central Europe in the Late Bronze Age and through the Iron Age (c. 700 BCE to c. 400 CE)."[2]

Back to our story.

When All Hell Broke Loose, but at the Home of the Neighbors

Things got really weird—totally ominous, one might say—when Leslie and Colin decided to take the two stones into the family home. That, unfortunately, proved to be a mistake; a *very big* mistake. It was a near-irreversible mistake, too. The story, and the tumultuous events that are very soon to come, weren't going away at any time soon. It was a case of opening something akin to the legendary Pandora's box. No one should be pondering doing something along those lines. The kids, though, did precisely that.

Very intriguing, and relative to the theme of this book, is the fact that those small stone heads began to move around the Robson home. Talk about creepy! And of their very own volition, too. Items in the house were inexplicably broken, in violent fashion, too. A glass was shattered while no one was present. Clearly, the stones had an ability to create havoc and mayhem—and to create sensational activity, just like what was seen at the Rollright Stones, as you'll see imminently. The next development took place not at the home of the boys, but at the home of their immediate neighbors. And it all began when the other kids in the street took the stones into that neighboring home. They did so in the dead of night, of course. Such situations like this one seldom go down on bright and sunny days.

Chaos in the Street

The next-door neighbor was a woman named Ellen Dodd. It wasn't long after the heads were taken into her home that unbridled chaos began. And it certainly didn't stop. Dodd was up and wide awake in the early hours of one night with her daughter, who had a toothache, when she had a terrifying encounter with something absolutely monstrous. Both Dodd and her daughter were confronted by nothing less than what looked like a hair-covered, large werewolf or, in their words, "half-man, half-beast."[3] It's hardly surprising that both mother and daughter were plunged into states of utter fear. That supernatural Hexham wolf of 1904/1905 was back, after so many years.

Mr. Dodd raced into the room where daughter and mother had quickly been plunged into a combined state of hysteria. This was no dream or hallucination: They could hear the monster padding down the stairs as if on its hind legs, as they worded it later. When the family finally dared to check out the rest of the house, they saw the front door was wide open. Had the marauding thing charged out of the door and into the darkness of the early hours? That was their total belief. Now, the story becomes even stranger, if such a thing is possible.

Ancient Folklore, Carved Heads, and a Terrified Doctor

It didn't take long for the story to get out to the local media. From there, the national media was soon screaming to get all the information, too. The saga of the Hexham Heads even made it on the BBC's prime-time TV show of that particular era, *Nationwide.* One person who happened to have heard of the story, who saw the *Nationwide* show, and who was determined to try to figure the truth of the eerie situation was Dr. Anne Ross. She held a PhD in archaeology and was an expert on the world and lives of the Celts. Not only that, Dr. Ross was a consultant for the National Geographic Society. Dr. Ross wrote a number of books, including *Folklore of Wales, Druids: Preachers of Immortality,* and *Folklore of the Scottish Highlands.*

Dr. Ross had a very good reason why she dearly wanted to see that unsettling pair of stones for herself—and to see them very quickly: She had some near-identical carved heads of her own at her home. Some of those in her collection were close to two thousand years old. The doctor was sure that the two carved heads that the Robson kids had found in their yard amounted to a very important part of ancient Britain and its heritage. Thanks to a few phone calls, it wasn't long before the heads were in Dr. Ross's hands, although, the time would come when the doctor wished she had stayed away from the entire, hellish, spooky situation.

At the time when the chaotic Hexham saga was swirling around, Dr. Ross was living in the south of England, in the city of Southampton, which was a long way away from Hexham in northern England. Nevertheless, Dr. Ross was determined to get the heads in her hands, one way or another. When she did so, however, something else happened. It was something that was just as terrifying and as bizarre as the events that the Dodd family were plunged into. It was only days after the doctor got hold of both heads and headed back to Southampton. So the rumor goes, Dr. Ross saw one of the stones, the female one, floating around the living room for a few seconds, before dropping to the floor in the living room. Reportedly, the doctor kept that part of the story hidden for many a year. That's hardly surprising for someone with such a well-respected background and character.

Then, on the very next night, and well into the early hours, Dr. Ross was violently woken up by the horrific sight of a hair-covered humanoid. It had a muzzle. And pointed ears. It was, then, something very much akin to the traditional werewolf of folklore. Or, it was the Hexham wolf resurrected, as some researchers, and locals in the town, suggested further down the line. Worse still, the hideous creature was in Dr. Ross's bedroom, looming and leering from above. Terror broke out in the Ross home, just as it had done at the Dodd home.

Paralyzed with Terror

Another rumor—not substantiated, I should stress—was that there had been some kind of violent, brief, sexual aspect to the doctor's nighttime encounter; something known as "sleep paralysis." Its title is a perfect one. If you have ever woken up around 3:00 a.m., in bed, unable to move, and with a dangerous, hideous creature standing by your bed (or, worse still, *on* your bed) that's sleep paralysis.

Writing for WebMD, Beth Roybal says of this genuinely terrifying phenomenon:

> *Over the centuries, symptoms of sleep paralysis have been described in many ways and often attributed to an "evil" presence: unseen night demons in ancient times, the old hag in Shakespeare's Romeo and Juliet, and alien abductors. Almost every culture throughout history has had stories of shadowy evil creatures that terrify helpless humans at night. People have long*

sought explanations for this mysterious sleep-time paralysis and the accompanying feelings of terror.[4]

Sleep paralysis could be considered a reasonable explanation for Dr. Ross's encounter. Some of the aspects are certainly similar. The doctor was, however, absolutely certain that, at the time of the face-to-face incident with that closest thing to a werewolf, she was 100-percent wide awake.

Parallels with the Dragon Project

Dr. Ross would later go on to say of the beast itself:

> *It was about six feet high, slightly stooping, and it was black, against the white door, and it was half animal and half man. The upper part, I would have said, was a wolf, and the lower part was human and, I would have again said, that it was covered with a kind of black, very dark fur. It went out and I just saw it clearly, and then it disappeared, and something made me run after it, a thing I wouldn't normally have done, but I felt compelled to run after it. I got out of bed and I ran, and I could hear it going down the stairs; then it disappeared toward the back of the house.*[5]

The Story Gets Bigger and Bigger

The graphic and detailed statement that Dr. Ross made suggests that she *did* have an encounter with something that was *not* born in the heart of her mind and her subconscious. This was all extremely similar to what had happened with the Robson and the Dodd families previously. And it all revolved around ancient, carved stones that moved around.

Later, it would become abundantly clear that some of this activity—sightings of hairy monsters and of strange stones—paralleled some of the sinister things that went down at the Rollright Stones, back in 1977 when the Dragon Project was hard at work, as you will recall. Things got worse for Dr. Ross and her daughter, Berenice: The Ross family did not just have one encounter with that hair-covered abomination. The creature appeared on several *other* occasions in the family home, as it did with the Dodd family miles away in Hexham and in the middle of the night. The creature was clearly determined to create mayhem and terror for everyone. It worked. Dr. Ross was absolutely certain there was an "evil presence"[6] in the family home, to the point where she finally decided to get rid of the pair of stones. By now, enough really *was* enough.

The Mystery Ends

After the Rosses wisely decided to get rid of the two ancient heads, the stones made more than a few trips, one might say. For a brief period they were held, and studied carefully, by staff at the British Museum who were experts in Celtic lore. Later, the heads were handed over to one Don Robins. You'll remember him: He was the author of the 1985 book *Circles of Silence* and was connected to the Dragon Project. Indeed, much of Robins's book was driven by matters relative to "the mystery of stone,"[7] and to whether or not there was "any truth in the belief that energy emanated from the stones, not only at Rollright," but at just about every ancient site throughout the British Isles."[8] No wonder Robins wanted to get his hands on the Hexham Heads. Those stones were not destined to stay with Robins, however. You probably know why: They creeped the hell out of him.

The list of people who got their claws into the stone heads of Hexham continued to grow. A dowser named Frank Hyde took control of them for a while. He, too, got a weird feeling whenever the heads were around him. He did what he felt was the right thing: He got rid of the infernal, paranormal pair. Today, the location and the ownership of the heads are curiously unknown. The story is unlikely to go away, though. For example, in 2010, Paul Screeton, an English journalist and author, wrote a 256-page book on the mystery titled *Quest for the Hexham Heads* that makes for fascinating reading. It regenerated the story and brought additional data to the fore. Despite all of his work into the story, Screeton is perfectly fine and has not been the victim of a wild, werewolf attack. Not yet, anyway.

"Energy Generated by Traumatic, Emotional, or Tragic Events"

With all of that said, it's now time to reveal how and why all of these bizarre angles—of Celtic heads on the move, of ancient stones, and of images of monsters seen after sunset—all come together as one. Also, it's time to see how all of this connects to ancient levitation in various sites across the planet. It's important to note there is a fascinating theory to explain all of the above, and much more, too. And it demonstrates the almost magical powers of certain stones. It concerns something termed "The Stone Tape," a phenomenon that dates back decades. It just so happens, however, *The Stone Tape* was *also* the title of a much-revered BBC production of the 1970s. Both the fiction and the fact of all this amounts to two parts of one eye-opening scenario. We begin with the world of television and entertainment.

At every Christmas in the UK in the 1970s, the BBC broadcast a creepy tale under the banner of *A Ghost Story at Christmas*. All of the stories were incredibly popular. *The Stone Tape* was particularly so. Penned by a playwright named Nigel Kneale, the story begins with a group of scientists, of a company called Ryan Electrics, who move into a large, atmospheric 19th-century building. As the story progresses, it becomes very clear that the old mansion has ghosts within its walls. These ghosts are not of the traditional type, however. They are much stranger than that.

"Typically Stone, Ideally Quartz or Limestone"

We're not talking about the dead coming back to life. Not at all. Rather, as the story progresses, it becomes clear that the ghostly phenomena, which is the theme of the story, is nothing of the sort. What the team of scientists are *really* encountering—and what, specifically, a computer programmer named Jill Greeley sees over and over again—are what we might term replays or recordings, or, even more specific, a never-ending loop that can only be seen by particularly sensitive people. Most important of all, there's a fascinating connection to these creepy images in the show: stone. Now, let's leave the BBC behind and look at the "Stone Tape" phenomenon in the *real* world.

The Ghost in My Machine says: "According to the stone tape theory, the energy generated by traumatic, emotional, or tragic events imprints records of those events on nearby physical objects—*typically stone, ideally quartz or limestone* [italics mine]—in much the same way that magnetic tape recording imprints data on strips of magnetic tape."[9]

The Stone Tape Theory website has addressed this issue, too: "By the late 20th century, belief in the Stone Tape Theory or residual haunting was commonplace amongst paranormal investigators. Many pointed to ancient cultural beliefs about spiritual stones and the importance numerous cultures have placed on certain objects in landscapes."[10]

Back to the Horrors of Hexham

There's yet another angle to all of this extremely strange controversy: that, perhaps, in ancient times, those who had a deep understanding of the nature of stone, of sound, of supernatural "loops," and of their paranormal potentials were able to create vivid and horrific images of werewolf-type beasts, Bigfoot-like things, and other monstrous entities—and where stone existed in abundance. In essence, it all boils down to this: Ancient tribes, centuries ago, protected their land—and particularly protected their ancient stone circles—from enemy tribes, and they did so in a very alternative way: by unleashing on their foes what I term *supernatural guard-dogs,* which were really nothing other than ancient equivalents of those aforementioned loops. One of those may very well have been the unholy beast that Dr. Anne Ross encountered back in the 1970s and that almost terrified her to death. The very same can be said for Ellen Dodd and her daughter.

That might sound outlandish to some people. The fact is, though, that such an intriguing scenario would go a long way to explaining how, and why, two young boys, in the north of England, in 1972, were able to stumble onto a pair of ancient stone Celtic heads, take them into the family home, and then, not long afterward, have the next-door neighbors get a hideous visit from a terrifying, rampaging werewolf. That would also go a long way to explaining how Dr. Anne Ross encountered a werewolf-like thing when she, too, handled the Hexham Heads. And it would explain the presence of that mysterious, hairy beast seen briefly in the heart of the Rollright Stones in 1977. I should stress that there are countless such cases—of monsters seen at ancient sites—and all of them are tied to ancient stone circles of varying types. Let's take a look at some of these chillingly similar situations to the events that went down at Hexham and that scared Dr. Anne Ross out of her skin. You may be surprised by their similarities.

The Weird World of the Nine Ladies and Their Resident Monster

In November 1991, a large, hairy creature—very much like the huge, hairy Bigfoot of the United States—was seen by an entire family at the Ladybower Reservoir in an area of the UK called the Peak District. Of course, the very idea that colonies of Bigfoot creatures could live in a nation as small as the UK, and *never, ever* get caught by anyone, is absolutely ridiculous. Yet, these rogue cases abound from the top of Scotland to the bottom of England. What is particularly notable about the 1991 encounter is that it happened in less than one mile from a place called Stanton Moor. It's a place not at all too different from Bodmin Moor, a huge area of land that we discussed earlier in these pages. And, as you can probably guess now, Stanton Moor is the home of a standing circle of old stones called the Nine Ladies. Despite the name, there are actually 10 of them; the final one was found and excavated in 1977.

The Nine Ladies circle and a strange creature (Wikimedia Commons)

Before we get to the monstrous encounter itself, here's some fascinating information on the Nine Ladies. *English Heritage* provides some interesting information on the old stones:

> *Stanton Moor is situated on elevated ground to the west of the River Derwent, near Bakewell in the Peak District. Few of the thousands of visitors who enjoy the tranquility of the moor and the fine views can fail to notice the widespread archaeological remains that are dotted across this landscape. Most of these are thought to date from the Bronze Age, about 3,000 to 4,000 years ago.*[11]

Let's Go Peak District says something about the Nine Ladies stone circle that will no doubt resonate with all those reading this book: "It gets its name from a legend that nine ladies were turned to stone as a punishment for dancing on the Sabbath, with the tenth stone, or King Stone, being the fiddler."[12] There's something that I suspect you've already anticipated: On occasion, the stones of the Nine Ladies are said to have been seen "dancing" on dark nights on the nearby ancient landscape. And, get this: to the eerie sound of a fiddle. Again we have stones moving and music present. A distortion of acoustic levitation? What else could it be? As I see it: nothing else, that's what. This brings us to back to that 1991 affair.

The Words of a Monster Hunter

Jonathan Downes is a longtime UK-based cryptozoologist—a "monster-hunter," in simpler terms—who has investigated a huge number of cases concerning strange and unknown animals in the United Kingdom. One of those cases took place *extremely* near to the incident at Stanton Moor. Downes stated that the family

> *brought the car to a screeching halt and came face to face with an enormous creature about eight feet tall, that was covered in long brown hair with eyes just like a man's. Its walk was different, too, almost crouching. But just as the man-beast reached the road, another car pulled up behind the family and blasted their horn—apparently wondering why they had stopped in the middle of the road. Suddenly, the creature—which I presume was startled by the noise—ran across the road, jumped over a wall that had a ten-foot drop on the other side, and ran off, disappearing into the woods. Now, I know that the family has returned to the area but has seen nothing since.*[13]

Back to Stonehenge

Merrily Harpur is the author of a fascinating book titled *Mystery Big Cats*, a book that addresses some of the strangest creatures seen in the UK: large, dangerous, black cats of the leopard type and size. Clearly, such fanged fiends should not be roaming and prowling anywhere around the country. Yet, the stark fact is that people *do* see them on disturbingly frequent times, and not only in the wilds of the UK, but within the hearts of the cities, too. As well as having chased down many such reports of what are known as "Alien Big Cats," or "ABCs," Harpur has tackled a case not unlike the one that took place at the Nine Ladies stone circle in 1991. Harpur's source for this particular story was a man named George Price. He was a member of the British Army and at the time (2002) he was taking part in a training operation on Salisbury Plain, on which Stonehenge itself just happens to stand.

At the height of the military exercise, a strange and horrific creature appeared on the landscape. Price described it as a huge ape that "seemed to be terrified due to the noise of all the military vehicles, such as tanks and jeeps."[14] Somewhat Orang-Utan-like, the creature raced farther away into the wilds of the massive Salisbury Plain that covers close to a huge three hundred square miles.

Scotland's Most Famous Monster Gets on the Scene

We now come to two more stories that strongly suggest we're on the right track, that of old stones formations, monsters, and guardians that really aren't genuine monstrous animals. They are millennia-old loops designed to protect their land. In their case, "land" amounts to stone circles. Another part of this story is connected to one of the most famous, and legendary, of all monsters: Nessie. That's right: the Loch Ness Monster. Over the years, all manner of theories have been put forward for what the Nessies might be. Plesiosaurs, marine reptiles that existed during the same time when the dinosaurs were on the verge of extinction, are one theory. Giant eels have been suggested by a lot of cryptozoologists. Massive salamanders have been offered.

The fact is, though, there is something very strange about the Nessies—something that suggests they, like all of those monsters seen at numerous stone circles in the UK, are also paranormal in nature, rather than being flesh-and-blood in nature.

The Bizarre Tale of the Stonehenge of Loch Ness

In 2007, the now-late conspiracy-theorist Jim Marrs wrote a book titled *Psi Spies: The True Story of America's Psychic Warfare Program.* It was a study of how, and why, the U.S. government, the CIA, and other agencies used the power of the human mind to try and penetrate some of the

Russians' most significant secrets. China's, too. Today? Probably North Korea and China. This powerful and paranormal ability is popularly called remote-viewing (RV). We have seen it before, in relation to the 1950s secret remote-viewing of ancient Egypt by the U.S. government.

Not only that, and for reasons that were never made clear, in the mid-1980s, the government's remote-viewers decided to try and RV none other than the Nessies, the world's most famous lake monsters. The CIA's psychics, however, got something they didn't anticipate: a creature that wasn't entirely real—at least, not in the way they expected.[15]

Astonishingly, Marrs said that although the Nessies could quite easily be picked up on sonar, something that suggested they were flesh-and-blood animals in nature, they also seemed to have the ability to vanish into nothingness. As in literally. It was as if, the CIA carefully recorded, that the Loch Ness monsters were nothing less than ghosts. Or, maybe of a paranormal or supernatural nature. At least, that was the best explanation that the CIA could come up with. At this point you might think, *So what? What does this have to do with the cases we addressed earlier in this chapter and relative to old stone formations?* Well, the answer to that is the most important part of this story: It's a little-known issue that Loch Ness has *its* very own Stonehenge. That's right: *yet another* saga of *yet another* beast that, incredibly, was not quite corporeal and whose presence in the legendary, 22-miles-long loch was attached to ancient stones.

Looking for Monsters, but Stumbling over Stones

The next part of the story takes us back to the summer of 1976 and revolves around a man named Marty Klein. At the time, Klein was out at Loch Ness, its water almost black in color at all times and with a depth of 745 feet. Like so many others who are fascinated by the mystery of the Loch Ness Monster, Klein was at the loch searching for monsters. He was part of a program ran by a Dr. Robert H. Rines. He was a man who had a deep passion for the quest to solve the Nessie riddle. The whole thing was funded by Boston's Academy of Applied Science. While the team was hard at work underwater, and wondering if a monster or several just might be caught on camera, they uncovered something amazing. Near one part of the loch, specifically the village of Lochend, Klein had one of those "I couldn't believe my eyes" moments. No, Klein did not see a Nessie or several. What he *did* see, though, was something totally different, but, equally, amazing. It was not just one, but a pair of structures around 30 feet below the surface.

The memorable, collective wording of the group was that they had hit on "a mother lode of stone circles—big ones and little ones; single circles and circles intertwined with others; circles laid out in a straight line and circles in no particular order."[16] They quickly became known as Klein-Henge One and Klein-Henge Two. While some of those stone circles were small—their diameters barely more than 10 feet—others were about 150 feet in diameter. Preliminary checks suggested that the circles stretched back to no less than Neolithic times. This was a major discovery. For us, it's another thread in the story of the mysteries of ancient stone. Let's learn a bit more about those monsters and their attendant stone circles.

"A Hugely Significant and Exceptionally Well Preserved Prehistoric Site"

In October 2018, I gave a lecture at the Mahnomen, Minnesota Casino and Hotel at the ParaCon Paranormal Convention. The lecture was on the stranger side of Nessie, including that story of the Loch Ness and the CIA. I said to the audience:

> *When the story of the stone circles of Loch Ness got out, Klein and a colleague, Charles Finkelstein, told the Chicago Tribune's John Noble Wilford that while a great deal of further research was required, it was their initial conclusion that the formations were fashioned by human hand and that they quite possibly dated back thousands of years, and to a time when the waters of the loch were at much lower levels than today.*

What that means, of course, is that at one point the water level at the loch, in ancient times, was far less than it is now. Thus, that allowed the people of that era to create their very own stone formations.

Now, say hello to the Clava Cairns. From *Visit Scotland*, we have the following in relation to the cairns:

> *Clava Cairns are a well-preserved Bronze Age cemetery complex of passage graves; ring cairns, kerb cairns and standing stones in a beautiful setting. Clava Cairns or the Prehistoric Burial Cairns of Bulnuaran of Clava are a group of three Bronze Age cairns located near Inverness. A hugely significant and exceptionally well preserved prehistoric site, Clava Cairns is a fantastic example of the distant history of Highland Scotland, dating back about 4,000 years.*[17]

The ancient Clava Cairns are situated not at all far from Loch Ness.

A Battle, a Beast, and Spinning Stones

The above data brings us to the Battle of Culloden, which was one of the most significant confrontations in Scottish history. Writer Ellen Castelow notes that the confrontation was the "last ever pitched battle to be fought on British soil [and] took place on 16th April 1746 on Drummossie Moor, overlooking Inverness."[18]

Now it gets really interesting. On the night of the tumultuous battle, a huge monster that was part-bird and part-bat, the Skree, circled ominously above and over and over again. Just about everyone knew of its terrible presence, and none could fail to hear its mighty, flapping wings. They were both signs of an awful omen: The Jacobites lost more than 1,200 men. The battle was over. The omen was right.

Consider this: We have two monsters in Scotland. One is a Nessie that, as the U.S. government's remote-viewing team suggested, was ghost-like, and another one that was a flying beast and that could easily rival the legendary winged Mothman of Point Pleasant, West Virginia, in the battle stakes. We also have stone circles below the surface of Loch Ness, and at the Clava Cairns, a Bronze Age cemetery. The distance from Loch Ness to Culloden is fewer than 20 miles. Connections, here, absolutely abound. How many more times do we need to be told of the connections between monsters and ancient stone circles? I'd say that, by now, we've got all that we need on this issue of monsters and stones.

There is, however, one final, key point: In 1987, there was a major expedition to try and solve the mystery of the Loch Ness Monster, once and for all. It was called Operation Deepscan. In the same lecture previously mentioned, at ParaCon 2018, I said the following:

> *Operation Deepscan was a highly ambitious October 1987 effort to seek out the Nessies with sonar and which just may have had some degree of success. No less than two dozen boats were utilized to scan the depths of Loch Ness with echo-sounding equipment. Some of the presumed anomalies recorded were actually nothing stranger than tree-stumps. Others may have been a seal or two, which had wandered into the loch. Nevertheless, there was a moment of excitement when something large and unidentified was tracked near Urquhart Bay and at roughly 600 feet below the surface.*

The whole thing prompted Darrell Lowrance, of Lowrance Electronics, whose echo-sounding equipment was used in Operation Deepscan, to say: "There's something here that we don't understand; and there's something here that's larger than a fish, maybe some species that hasn't been detected before. I don't know."[19]

There's one more part to this story, a very strange part. At the exact same time that *Operation Deepscan* was doing its very own thing, a man named Vic McDonald was visiting the Clava Cairns. Something very strange happened to McDonald while he was on his vacation. He said, that, to his astonishment, he very briefly saw what he termed a "procession" of small stones—all of them around the size of a football—"spinning" above what are termed the three cairns at Balnuaran of Clava. In seconds, they shimmered and vanished.[20]

How about a story, next, of the ways and means by which the ancients used sound and stones in an even stranger fashion? The bulk of this story revolves around acoustic levitation. There is *another* aspect of sound that plays a role in all of this. It's called infrasound. It's that part of the saga that provides the answer to why so many people can see so many of those "supernatural guard dogs," or, more correctly, "guard monsters." As we'll see in the next chapter, the Levitators clearly mastered both phenomena: levitation *and* infrasound. Put the two together, as we surely will imminently, and you'll get to see how all of this comes together and in one big, amazing picture—with the answers provided.

CHAPTER 22

ANOTHER KIND OF SOUND COMES FORWARD

In this chapter, we look at another form of acoustics, one that is the key to understanding those mysterious monsters seen at multiple sacred stone circles, and largely in the United Kingdom (most notably at the Rollright Stones). It demonstrates the incredible extent to which sound, in varying ways, meant so much to the Levitators. They were not just master architects when it came to acoustic levitation; they were also a race that had conquered other aspects of sound, and for multiple reasons, as will become apparent. That brings us to the matter of what is termed as infrasound. It's a phenomenon that can have significant effects on both the human body and mind. Mostly, exposure to infrasound has an adverse result on people. We'll get the ball rolling with *Audio Visual*. They provide us with a good, solid summary of the subject:

> *The simplest definition of Infrasound, sometimes referred to as low-frequency sound, is sound that is lower in frequency than 20 Hz or cycles per second, the "normal" limit of human hearing.*

Hearing becomes gradually less sensitive as frequency decreases, so for humans to perceive infrasound, the sound pressure must be sufficiently high. The ear is the primary organ for sensing infrasound, but at higher intensities it is possible to feel infrasound vibrations in various parts of the body.[1]

It's important to note there is far more to infrasound than just that. It is something that has a connection—in a fascinating fashion—in the animal kingdom, too.

Government Interest in Infrasound

The most visible data on the connection between animals and infrasound comes from the U.S. government's National Oceanic and Atmospheric Administration (NOAA), which has studied infrasound deeply and states:

Whales are very social creatures that travel in groups called "pods." They use a variety of noises to communicate and socialize with each other. The three main types of sounds made by whales are clicks, whistles, and pulsed calls.

Clicks are believed to be for navigation and identifying physical surroundings. When the sound waves bounce off of an object, they return to the whale, allowing the whale to identify the shape of the object. Clicks can even help to differentiate between friendly creatures and predators. Clicks have also been observed during social interactions, suggesting they may also have a communicative function. Whistles and pulsed calls are used during social activities. Pulsed calls are more frequent and sound like squeaks, screams, and squawks to the human ear. Differing vocal "dialects" have been found to exist between different pods within the same whale population. This is most likely so that whales can differentiate between whales within their pods and strangers.[2]

Mick Hamer, at *New Scientist,* gives us this:

> *Tigers appear to rely on booming low-frequency sounds—much of it inaudible to humans—to drive rivals away from their territory and to attract mates. The discovery may explain how the animals maintain large hunting territories, and may also help conservationists to protect the endangered animals. Tigers produce a wide variety of sounds, from deep roars and growls to the raspberry-like "chuffing" they use to greet each other. A roar followed by a growl is probably designed to intimidate rivals.*[3]

Acoustics.org have even more fascinating material for us to work with:

> *In 2000 it was discovered that tigers, like whales, elephants, rhinos, and other animals, could create sounds low in frequency and some that are infrasonic and inaudible to the human ear. Twenty-four tigers were recorded at the Carnivore Preservation Trust in Pittsboro, N.C., and the Riverbanks Zoological Park in Columbia, S.C. It was found that tigers can create sounds around 18 hertz.*[4]

It's now that we come to the really strange part of all this and how it plays in with the overall theme of this book.

"A Sense of Awe in Congregations"

In 2003, Reuters put out a release on infrasound that got picked up by numerous news outlets. For us, at least, here are the most relevant parts of the article:

> *Mysteriously snuffed out candles, weird sensations and shivers down the spine may not be due to the presence of ghosts in haunted houses but to very low frequency sound that is inaudible to humans. British scientists have shown in a controlled experiment*

that the extreme bass sound known as infrasound produces a range of bizarre effects in people including anxiety, extreme sorrow and chills—supporting popular suggestions of a link between infrasound and strange sensations.[5]

"Normally you can't hear it," stated Richard Lord, of the United Kingdom's National Physical Laboratory, an acoustic scientist employed at the National Physical Laboratory and who worked on the project to fully understand Infrasound.[6] Associated Press continued: "Some scientists have suggested that this level of sound may be present at some allegedly haunted sites and so cause people to have odd sensations that they attribute to a ghost—our findings support these ideas, added Professor Richard Wiseman, a psychologist working at the University of Hertfordshire in the U.K."[7]

Says Sarah Angliss, an engineer and a composer: "This is not a new phenomenon; church organ builders have been using infrasonic pipes for over 250 years to create a sense of awe in congregations."[8] As all of this demonstrates, Infrasound causes strange effects in the human mind and body. It may also create imagery of strange creatures and monsters, as we've seen. Not literally, but in the depths of the human brain.

Bigfoot and Infrasound: A Strange Connection

One of the most intriguing accounts concerning infrasound and a monstrous beast comes from a Bigfoot researcher by the name of Scott Carpenter. Of his face-to-face encounter with one of the legendary, hairy giants (at a site he prefers not to reveal), Carpenter said: "I am not an acoustic expert or a scientist. My findings are based on observation and common sense. I think that I was under the influence of infrasound during my encounter with the Bigfoot on April 30th, 2010. The Bigfoot manipulated my perception and sanitized my memory."[9]

There was more data from Carpenter: "Even more disturbing was the fact that I did not react to observing the Bigfoot. I had to have initially recognized what it was and where it was hiding. I made two attempts to zoom in on the Bigfoot and get a close up video. Sometime during this process I was subjected to the influence of infrasound and strongly influenced or 'brain washed' into walking off. It is almost like my memory was wiped clean and I was given instructions to leave and I did."[10]

Based on all what we know so far, we might ask ourselves: Did Carpenter actually see a real, flesh-and-blood Bigfoot? Or, did a blast of infrasound cause Carpenter to have a visionary-type experience—a vision that was incredibly lifelike and monstrous? This is very much a case of which came first: the chicken or the egg? We've seen how, for years, in the UK, people have seen Bigfoot-type beasts, almost always near ancient sites. We also know, from the Dragon Project of 1977 and of the "Stone Tape" phenomenon, that residual energy can linger—and linger for a very long time. Could something akin to the UK phenomena have played out too with Scott Carpenter? Either way, sound appears to have been the culprit. And, it's important to note that this is not the only report of its type.

Another Example of the "Sound of Bigfoot"

"Miss Squatcher" is a well-known figure in the Bigfoot research community. She had her very own experience that involved infrasound and Bigfoot. Of an encounter she had at Elbow Falls (Alberta, Canada) in the summer of 2013, she recalls:

> *I felt as if my chest was heavy, my breathing was shallow and I could hardly catch my breath. I stood up from examining the scat and scanned my surroundings, the sensation of my pulse pounding in my head. I saw nothing. I could feel panic setting in. I was on the edge of a full-blown panic attack and had the unrelenting feeling that I needed to leave the area, now.*
>
> *My anxiety was increasing and I shared this with the others. They were startled when I took out my compass, oriented myself in*

the direction we had come and started walking straight through the bush. "I have to leave, I don't feel right." Was there a Bigfoot nearby producing infrasonic waves to scare the heck out of us? We will never know for sure. But I can say with certainty that something out there made me feel more fear and panic than I have ever felt before.[11]

Werewolves Created out of Infrasound?

Bigfoot is not the only hair-covered, upright monster that lurks in the forests and woods of the United States. There is the Dogman, too. Its name is all too relevant: The creatures, like dogs, can stand on all-fours and, also, on occasion on their back limbs. With their muzzles and their pointed ears, they are surely the closest things one can think of when it comes to the controversial matter of werewolves. Interestingly, investigator-researcher-author Linda Godfrey has spent several decades pursuing the Dogman phenomenon; she has noted that many sightings of these infernal things, too, are *very* often seen near sacred sites.

Godfrey is the author of author of many books and the acknowledged expert on the Dogman mystery. Those books include *The Beast of Bray Road* and *The Michigan Dogman.* As Godfrey states, there are "staggeringly large numbers of precisely formed earth mounds that graced the landscape; some 20,000 or more in at least 3,000 locations."[12] Their purposes? The primary ones are burial mounds, tribal totem makers, and ceremonial centers. In other words, Godfrey has made a connection between horrific, dangerous creatures in the United States and their presence at ancient sites. This tells us something incredible: It's not just in the United Kingdom. That people have felt dizzy, nauseous, and even sometimes in a vacuum in the presence of the American Dogmen strongly implies that, also in these cases, there was an infrasound angle—an angle that involves Stone Tape residue. And mind-bending residue.

A few closing words on infrasound from writer Seth S. Horowitz, who says:

People don't usually think of infrasound as sound at all. You can hear very low-frequency sounds at levels above 88–100 dB down to a few cycles per second, but you can't get any tonal information out of it below about 20Hz—it mostly just feels like beating pressure waves. And like any other sound, if presented at levels above 140 dB, it is going to cause pain. But the primary effects of infrasound are not on your ears but on the rest of your body.[13]

CHAPTER 23

THE AVEBURY STONES: FACT, FICTION, AND A GOVERNMENT FILE

I have one more example of how sound has played a major role in protecting ancient circles by using creatures that actually weren't "real" as we perceive them. Again, it is all due to infrasound and how it can frazzle the human mind.

Just twenty-four miles away from legendary Stonehenge are the equally intriguing stones of the ancient village of Avebury. *The History Press* says of the Avebury stones: "Avebury is best known as the world's largest stone circle: a quarter of a mile across, it partly encompasses a village. However, the stone circle is just one element of the Avebury Complex. Scattered across the ritual landscape surrounding Avebury is a wealth of prehistoric monuments, spanning many thousands of years in their construction."[1]

In concise fashion, *English Heritage* gives us this: "The Avebury complex is one of the principal ceremonial sites of Neolithic Britain that we can visit today. It was built and altered over many centuries from about 2850 BC until about 2200 BC and is one of the largest, and undoubtedly the most complex, of Britain's surviving Neolithic henge monuments."[2]

There is something else, too: Avebury is both mysterious and magical. In just a few moments, you will see precisely why that is the case.

The Entertaining Saga of the Children of the Stones

It's intriguing to note that, in 1977, when the Dragon Project was in full-steam-ahead mode at the Rollright Stones, a highly entertaining UK television show was grabbing the attention of the children (and more than a few of the adults, too, I dare say) of the country. I can say that with complete certainty. How can I be so sure? I'll let you in on it: I was one of those many kids who raced home, after the school day was over, and watched the show. Its title: *Children of the Stones*. I was just eleven years old when the filming of the show went ahead in 1976 at none other than Avebury, Wiltshire.

I won't give away the story (since you can purchase it on DVD and watch it for yourself, or you can read the novel of the very same story). It's safe to say, however, that it's one of the most gripping, and spooky, shows ever made for kids. What I will say, though, is that in the show we have a scientist and his son, a creepy Aleister Crowley–type character who blends science with ancient magic, and a number of villagers who, bit by bit, are transformed into something similar to the Stepford Wives of the movie of the same name. And there are the numerous, magical—and dangerous—stones of the village of Milbury. Those stones have awesome powers that date back to ancient times. It's not long, though, before the past and the present blend together in mysterious fashion and the story reaches its peak.

For the filming, Avebury doubled for the fictional Milbury. For a pre-teen Nick Redfern, the show, which ran for seven episodes, was great entertainment. Having watched the show again while writing this book, I can safely say that it's *still* great entertainment. Whether or not, with the connection to the work of the Dragon Project at the same time, there was some kind of strange and highly relevant Zeitgeist in the air at the time, I cannot say for certain. What I *can* say is that for a period in the late 1970s, fact, fiction, magical stones, and alternative science all combined together in eerie, swirling fashion in the UK. Now, we come to the weirdest part of the story of Avebury. This part of the story has nothing to do with TV entertainment. This one is all real.

A Real *X-File* of the Highest Strangeness

Over the decades, government agencies all across the planet have investigated what we might term "paranormal phenomena." The FBI has studied so-called "cattle mutilations." The U.S. Air Force spent years investigating UFOs. The CIA has a file on nothing less than Noah's Ark. I could easily keep going on. But, let's stay with the story that is most relevant to both you and me.

The story remained steadfastly classified for decades, under the terms of the United Kingdom government's "Thirty Year Ruling." As its name suggests, it was a tedious official piece of government regulation that meant sensitive, government data had to be hidden from the public for at least three decades, and sometimes longer. Today, however—and since the year 2000—the UK government has had a Freedom of Information Act, effectively now negating the need for the Thirty Year Ruling. It is within those old, now-creased and yellowed 1960s-era files—and in the vaults of the National Archives—that the Avebury story can be found.

It was thanks to a man named Malcolm Lees that the story ever got to the surface. Lees, who died just a few years ago, was an employee of what, back in the 1960s, was called the Provost and Security Services (P&SS), the equivalent of the U.S. Air Force Office of Special Investigations. Its headquarters, in the 1960s, was the spacious Government Buildings, at Acton, England. That's where the Provost and Security Services carefully guarded all of their most important files and records. In simple terms, they were the elite of the Royal Air Force, the people whose job it was to investigate matters relative to national security issues: reeling in Russian spies, tracking down terrorists, and ensuring the safety of the UK.

The UK National Archives: a hub of secrets (Nick Redfern)

As the Story Begins, So Does the Terror

One could say that the story of Malcolm Lees most assuredly fell into that category of national security, although in a very strange way. It all happened in the latter part of September 1962.[3] And Lees never forgot it. No doubt, that was very much the same for the witness, too. Lee said that it wasn't at all unusual for the P&SS to investigate UFO cases. In fact, they received about 15 to 20 reports per year. Mostly, they were just cases of completely unexciting "lights in the sky"–type reports filed by members of the public, who had first contacted their local police, and who then handed things over to the P&SS for scrutiny. The bulk of the reports turned out to have involved nothing stranger than satellites, and aircraft lights, seen at night. Not this case, though. No way. This one was totally something else.

The witness was a woman who lived in the village of Avebury, was in her 40s, and was alone, aside from her faithful dog, Jeremy. Interestingly, as the file showed, she had a big interest in archaeology and ancient history. Most nights she would go out into the darkness, often at about 10:00 p.m. or later, to walk Jeremy around the ancient stones of Avebury, and "armed" with just a flashlight. On this case, however, she had left Jeremy at home. He had been playing outside all day and was wiped out, so she decided to let him stay curled up in his basket.

A Monster in the Stones

At roughly 10:30 p.m., and walking along the main stretch of stones, the woman was suddenly stopped by the incredible sight of a dazzling globe of light coming directly toward her. It was, maybe, a couple of feet in diameter at most, roughly 15 feet away from her, and bobbing up and down like a small boat on a windy lake. It's no surprise that the woman was frozen in her tracks; her eyes practically popped out of their sockets. Amazement turned to absolute terror when the ball of light came closer and "nearly exploded in front of me," and changed—shapeshifted, one might even say—into the form of a "horrible, big worm which wriggled right in front of me with his eyes."[4]

That the bright flash briefly affected the woman's ability to see the thing clearly, only created even more terror for the poor woman. The thing itself was about 5 feet in length, gray-white in color, and slightly less than a foot in width. It's hardly a surprise that the woman, fearful of there being more of the creatures lurking among the stones, fled from the scene and back to her home. She frantically called the police, raced to her bedroom, and locked the door. When morning came, the woman tentatively opened the front door, and with Jeremy for support, although, it's doubtful the mayhem had affected him. Thankfully, the thing was gone by morning. It wasn't long before a pair of Provost and Security Services personnel were driving into the village of Avebury to investigate one of the strangest UFO-like case they ever had heard of.

The Military Gets Interested

According to Malcolm Lee's memories, his colleagues assumed—at first—the whole thing was surely going to be just someone's idea of a big joke. When the two P&SS agents arrived, though, it quickly became clear that this was no joke. They were confronted by the woman, who was still in a state of terror. By the time the P&SS agents arrived, the worm/ball of light was nowhere to be seen. It was long gone. That was not a bad thing. Avebury looked as just atmospheric, entrancing, and tranquil as it always had been. It was almost as if the nightmare never took place—except for the fact that, for the woman, it was all too real. And all too horrific.

There was one additional angle to all of this. The two men asked the woman to go with them to the specific stone where she saw the writhing monster. Intriguingly, when the trio got to the site, there was a line, about 12 or 13 feet in length, of a silvery, slimy, sticky goo on the grass. One of the two agents asked the woman if he could borrow a glass and use it to scoop up some of the wormy slime. She was happy to help, now that she was calming down. The other man asked a bunch of questions to ensure the whole story had been chronicled.

Interviewing the Witness

I should stress that there were no Men in Black-type interrogations. There were no threats or warnings not to tell anyone what had happened the night before. In fact, after the slime was collected, and the area was carefully studied and photographed, all three sat down, drank tea, and ate homemade cakes, courtesy of the women herself. After they went over the story one more time, the P&SS men thanked the woman, and the two men went on their way. Another thing: When, four or five days later, the woman called a phone number the two men had given her if she needed to speak with them again, the response she got was that her glass had been lost—along with the slime. Years later, Malcolm Lees would go on to say that he felt that the issue of the glass and the slime vanishing seemed slightly suspicious. Hardly a conspiracy of "Who shot JFK?" proportions, but perhaps something slightly curious, still, to ponder on that day.

Strange Parallels

To go back a bit, we should also ponder on the fact that fourteen years after the August 1962 incident occurred at Avebury, the hugely popular UK-based television series *Children of the Stones* entertained the kids of the country for seven weeks. There are fascinating similarities between the two: *Children of the Stones* was filmed in Avebury, even though its name was changed to Milbury for the show. In the story as told by Malcolm Lees, the ball of light shapeshifted into a wormy monster. In the TV show, many of the villagers of the fictional village of Milbury were shape-shifted into stones. So, there's a case of changing form in both.

And, Malcolm Lees, after carefully watching *Children of the Stones* on several occasions, was certain that the stone closest to that worm-thing seen back in 1962 was actually filmed—deliberately or not—during the shooting of *Children of the Stones*. A bizarre series of strange synchronicities? Who knows? If nothing else, it's yet another near-all-blanketing layer of the weirdness that revolves around the magical, mysterious stones of our world.

Mysterious Lights and Stones that Can't Stand Still

Finally, on the matter of Avebury, there's something particularly fascinating. The team of Janet and Colin Bord, whom we heard of in chapter 8, spent a lot of time studying the history of the ancient village. They said that strange, paranormal activity occurred in and around the stones of the village for a long time. First, mysterious music had been heard around the Avebury stones—and *within* the stones, too. Second, the Bords said that weird lights had been encountered in the confines of the village.

They said: "During the First World War, Edith Olivier, author of books on Wiltshire, was driving through Avebury at twilight and heard the music and saw the lights of a fair among the stones. When she later remarked on this, she was told that it was at least fifty years since a fair had been held there."[5]

One of the legendary standing stones at Avebury, England (Nick Redfern)

Could those illuminations have been connected to that ball of light encountered at Avebury, back in August 1962? It's definitely something to wonder about. Equally fascinating, Edith Olivier, a well-known author who wrote two books on the county of Wiltshire, received a story, many, many years ago, of a local person who briefly saw one of the old stones "walking" across the village after midnight. Such a thing sounds completely bizarre, yet the same story has the identical templates we've seen on so many occasions: (a) stones that could move, and (b) sound in the form of music.

CHAPTER 24

THE DANGERS OF SOUND

Earlier in this book, we saw how, in an amazing fashion, the walls of the city of Jericho were brought to the ground by the use of sound. This is not new. It's a lesser-known fact that acoustic technology has been seen as a weapon for a very long time, rather than for construction. The Levitators knew all about it. Jericho is the perfect example. Disturbingly, much of this technology today is being directed at us (the public). That's right: For years top-secret experiments have been run to see just how effective sound can be when it is focused on the human body and the mind. One of the most significant examples of all this comes from the U.S. Defense Intelligence Agency (DIA). Relevant or not, it was the very same U.S. government agency that opened a file on airplane pilot and "UFO grid" proponent Bruce Cathie in the 1960s.

In March 1976 the DIA really got into the heart of sound and its potential as a weapon of war. For years, the document was classified and totally out of the hands of the public and the media. Its title is "Biological Effects of Electromagnetic Radiation (Radiowaves and Microwaves) Eurasian Communist Countries." It was penned by two men, R.L. Adams and R.A. Williams. Both men were in the employ of the U.S. Army's Medical Intelligence and Information Agency. The pair wrote: "The Eurasian Communist countries are actively involved in evaluation of the biological significance of radio-waves and microwaves. Most of the research being conducted involves animals or in vitro evaluations, but active programs of a retrospective nature designed to elucidate the effects on humans are also being conducted."[1]

There was something else: something seen as being almost impossible but that, incredibly, became absolutely possible.

The Nightmares of "Repeaters"

With a mixture of research into microwaves and varying frequencies of sound, the U.S. Army found, "thanks" to their top-secret experiments, people's minds could be altered in some of the most alarming ways, even to the point of "beaming" messages into the minds of their targeted individuals. Rumors later had it, although it's not mentioned in the document, that more than a few U.S. Army personnel suffered terribly from the short-term effects of hearing voices in their heads. That's hardly surprising; imagine having random voices in your head, and time and time again. What a nightmare. Most disturbing of all for the military was the fact that the experiments provoked what were termed "repeater" situations: people targeted by specific soundwaves in the experiments occasionally heard the very same voices in their heads, but sometimes days and weeks later. Imagine that: You're talking to friends, watching television, or driving your car, when, suddenly, you hear in your head the very same words that appeared in your mind days, weeks, or even months beforeand completely randomly. Thankfully, the sounds eventually went away. It was, though, a traumatic situation for those who agreed to go ahead with the tests.

The Russians Get Involved

It wasn't just voices that the military volunteers were plagued by. The Defense Intelligence Agency document of Adams and Williams reveals that other symptoms were "irritability, agitation, tension, drowsiness, sleeplessness, depression, anxiety, forgetfulness, and lack of concentration."[2] Imagine having all of those symptoms, and not even being able to control things in the slightest. The DIA report was made all the more disturbing because, while working in tandem with the CIA, it became clear that the then–Soviet Union was far ahead of the United States. That worrying fact made it into the DIA report.

The DIA said the following of the Russians' work:

> *Sounds and possibly even words which appear to be originating intra-cranially can be induced by signal modulation at very low average-power densities. The Soviets will continue to investigate the nature of internal sound perception. Their research will include studies on perceptual distortion and other psycho-physiological effects. The results of these investigations could have military applications if the Soviets develop methods for disrupting or disturbing human behavior.*[3]

Bear in mind, all of this work was undertaken way back in the mid-1970s. What about in latter days and the present day? Let's see.

Sound as a Weapon

Now we come to one of the most outrageous stories of the 1980s. It's a story that involved the loss of civil rights, the deliberate use of sound as a weapon, and zero care for the people who were caught in the maelstrom. It all revolved around Royal Air Firce Greenham Common, a closed-down military base in the UK. Here's some background on the base from the UK government:

> *Royal Air Force Greenham Common or RAF Greenham Common is a former Royal Air Force station in Berkshire, England. The airfield was southeast of Newbury, Berkshire, about 55 miles (89 km) west of London. Opened in 1942, it was used by both*

the Royal Air Force and United States Army Air Forces during the Second World War and the United States Air Force during the Cold War, also as a base for nuclear weapons. After the Cold War ended, it was closed in September 1992.[4]

Historic England expanded on the terrible story and the history of the base:

The first cruise missiles were delivered in November 1983 and by 1986 there were 96 missiles and five spares made up into six mobile cruise missile flights housed at GAMA [the Ground launched Cruise Missile Alert and Maintenance Area]. . . . Peace camps were established around the base perimeter fence and the Greenham women, in opposition to the deployment of cruise missiles, used non-violent protest to bring the nuclear capability of Greenham Common airbase and the campaign for nuclear disarmament to the attention of the world. It is for this reason that Greenham Common is a name which is internationally famous; a site which is symbolic of international anti-nuclear protest.[5]

This is when everything got very dark and disturbing for just about each and every one of those women who were only doing nothing but making their views known.

Blasting Women with Sound, *Supposedly* All in the Name of National Security

It became clear—very quickly so—that many of the women who were protesting outside the gates of RAF Greenham Common started to fall ill. The symptoms were not just alarming, they were *beyond* alarming: Short-term memory issues were the most reported symptoms to local doctors. Pummeling migraines crippled the women for hours. On occasions, those migraines hung around for days. Anxiety and full-blown panic attacks hit the women. Luckily, there were people on the base—confidantes and whistleblowers, one might very well say—who were in positions to quietly tell the women why and how, exactly, they were all suddenly falling sick. It turned out that sound was the reason. Luckily, the UK media were able to crack the outrageous situation and reveal the truth. It was specifically all thanks to the journalists of the UK's *Guardian* newspaper. The Air Force, however, was far from being happy.

Not long after the women got even sicker and sicker, the *Guardian* ran a feature in its pages with this eye-opening headline: "Peace Women Fear Electronic Zapping at Base." The article said that the base security personnel had in their possession what was described as "an intruder detection system called BISS, Base Installation Security System, which operates on a sufficiently high frequency to bounce radar waves off a human body moving in the vicinity of a perimeter fence."[6] Sound—and directed sound of varying frequencies—were at the forefront of the experiments targeted on all of the women.

Electromagnetic Activity around the Women's Camps

It is important to note that this story was not exclusive to the United Kingdom. The late conspiracy theorist Jim Keith quickly picked up on the *Guardian*'s article. He wrote: "A probable instance of the use of ELF [Extremely Low Frequency] weapons on civilians took place in January of 1985, when women protesting nuclear weapons at the U.S. Air Force Base at Greenham Common in England began experiencing strange physical symptoms, coincident with the changing of the security system at the base from primarily human to electronic."[7]

Keith also said of this outrageous affair:

> *Guards patrolling the base's perimeter were reduced in number, and antennae were installed at intervals. The women experienced symptoms characteristic of ELF irradiation, including headaches,*

earaches, pressure behind the eyes, bleeding from nose and gums, fatigue, the hearing of clicks and buzzes, and heart palpitations. Electronic testing performed by a Canadian scientist and a British electronics activist group confirmed distinct areas of electromagnetic activity around the women's camps.[8]

Today, the facility is no more:

On 11 September 1992, USAF returned RAF Greenham Common to the Ministry of Defense. In February 1993, the Greenham Common air base was declared surplus to requirements by the Secretary of State for Defense and the airfield was put up for sale. The Greenham Common Trust was formed in 1997 to run the technical side of the base which became a business park. The airfield side was opened to the public in 2000.[9]

When Humming Turns People Crazy

It may sound crazy or paranoid, but there are long-term rumors that the effects of sound have been directed on people all across the planet. The goal: to see just how, and to what degree, sound could affect the human mind and body. Certainly, the most controversial story concerns the Mexico-based "Taos Hum." Writer Ben Radford says:

The hum seems to have first been reported in the early 1990s. Joe Mullins, a professor emeritus of engineering at the University of New Mexico, conducted research into the Taos Hum. Based on a survey of residents, about 2 percent of the general population was believed to be "hearers," those who claimed to detect the hum. Sensitive equipment was set up in the homes of several of the "hearers," measuring sounds and vibrations but after extensive testing nothing unusual was detected.[10]

Astonishing Legends has got their grips into the story, too, such is its interest and concern: "No one really knows when or why the hum began. But, approximately 30 years ago, the residents in Taos, New Mexico noticed the constant high frequency humming noises in their town. What are they? Why has it been heard for over three decades?[11]

Finally, for the Taos Hum, there are these words: "A survey of the residents of Taos indicated that at least two percent of the population could hear the hum. Of the approximately 8,000 residents of the area, 1,400 responded to the survey, and 161 of the respondents claimed to hear the hum."[12]

The hum, unfortunately, continues to annoy. At least it's not killing.

Secret Military Activity

Moving back to the UK, there's the "Bristol Hum" that's titled after the English city of Bristol. The BBC said in 2016 of its equivalent, the Taos Hum:

> *Some people in Bristol say they are plagued by a mysterious low-level hum that no-one can trace. . . . But it's not the first time the hum has kept Bristol awake. In the 1970s hundreds of the city's residents complained to the council that a strange noise was audible at night. Most of the experts drafted in put it down to factory noise, electricity pylons or tinnitus—while some of the more imaginative suggestions included the sound from flying saucers hovering over the city or secret military activity. Eventually, it stopped as abruptly as it began, but not before it had spawned reports of equally unidentified hums in other towns across Britain.*[13]

"Future Riot Shields Will Suffocate Protestors with Low Frequency Speakers"

That, directly above, was the headline of a *Gizmodo* article that appeared on the internet in 2011. And it all had its connections with Raytheon. The company states of itself: "Raytheon Technologies Corporation is an aerospace and defense company that provides advanced systems and services for commercial, military and government customers worldwide."[14]

As for the Raytheon story that *Gizmodo* hit on, it concerned the controversial plans of the company to put into place nothing less than crowd-control technologies—and using sound to disable those seen as demonstrators. As an aside, what's wrong with demonstrating? Nothing, that's what. It should be encouraged.

Beware of the "Sound Cannon"

Andrew Liszewski, who wrote the feature for *Gizmodo*, said: "It's not the first crowd control tool to use sound waves, but Raytheon's patent for a new type of riot shield that produces low frequency sound waves to disrupt the respiratory tract and hinder breathing, sounds a little scary."[15] Just like some of the other technologies described above, Raytheon's would have the ability to provoke migraines, nausea, and vomiting. In simple terms: If you're hit by what's known as the LRAD Sound Cannon, you'll be disabled in astonishingly quick time.

And what, exactly, might that particular "cannon" be? *Pitchfork* says: "The Long Range Acoustic Device, or LRAD, is a speaker system and sound energy weapon developed in the early 2000s for use by the U.S. military. . . . LRADs have a microphone for speech, inputs for playback of recordings, and a built-in 'deterrent tone.' based on frequencies that are especially painful to the human ear."[16]

As all of this tells us, sound is one of the most fascinating phenomena on the planet and, as we have seen, it has played a significant role in world history. Talking of world history, we are now about to address one of the most important aspects of this overall story: Were the mysterious Levitators from another world, or were they an equally enigmatic group of people from here?

CHAPTER 25

A CASE FOR EXTRATERRESTRIALS IN THE PAST

Having taken a good look at the world's most amazing structures of massive proportions, it's time to address the most important angle of all: Who really were the creators of those gigantic statues, buildings, structures, and more? The Levitators were either humans or nonhumans. There really are no other options. Let's see if we can figure it out. We begin with the extraterrestrial theory. There's no doubt at all, as we've seen, that over the decades more than a few people have favored—championed, even—this particular angle. Indeed, the extraterrestrial (ET) theory does have its merits.

The Man Who Turned Anunnaki Gods into Rock Stars in Today's World of Ancient Extraterrestrials

Lawrence Gardner's words regarding the Anunnaki, which we heard of earlier in the book, were certainly vague, and even slightly evasive, in some ways. Zechariah Sitchin—the one person who championed and highlighted the Anunnaki phenomenon more than anyone else, and someone who studied economics at the University of London—was sure of his thoughts and theories of the alien kind. Sitchin, who died in 2010, came right to the heart of it all.

In his many books, which include *The Cosmic Code, The Wars of Gods and Men,* and *Divine Encounters,* Sitchin came to the undeniably controversial conclusion that the Anunnaki were not deities, demons, or supernatural entities of other types. No. Sitchin was sure that the Anunnaki were a race of incredibly ancient extraterrestrials who came from a world titled Nibiru. Not only that, said Sitchin, Nibiru was said to have been an elusive world that *still* orbits our solar system, albeit at a mind-boggling distance from our Earth.[1]

On top of that, Nibiru supposedly comes perilously close to the Earth every 3,600 years, provoking planet-wide disaster to our world, as a result of its incredible gravitational pull. Hence the world's ancient legends of countries pummeled and flattened, gigantic floods overwhelming the land, and nations totally destroyed.

Supposedly, too, one of the reasons why the Anunnaki came to Earth was for our gold. Of this part of the strange story, *Token Rock* says:

> *Throughout history, alchemists have sought the elusive Philosopher's Stone, the secret White Powder Gold which would become quite literally a vessel of the "light of life." This secret material was reported to bestow powers of immortality in addition to incredible supernatural powers to those who consumed it. Certain famous mystics, magicians, and alchemists of history like Enoch, Thoth, and Hermes Trismigestus are known to have perfected the sacred art of creating The Philosophers Stone, and their use of the material explains the many legendary supernatural powers ascribed to the White Powder of Gold (ORME).*[1]

Sitchin's Fascination with Baalbek and Those Gigantic Stones

For Sitchin, many of the huge stone creations that can be found all across our world were the work of the Anunnaki. For example, Sitchin was sure that there was a connection between alien technology and the huge stones at Baalbek. His quote, again:

> *The enigmas surrounding the site and the colossal stone blocks do not include one puzzle—where were those stone blocks quarried; because at a stone quarry about two miles away from the site, one of those 1,100-ton blocks is still there—its quarrying unfinished. . . . The quarry is in a valley, a couple of miles from the site of the "ruins." This means that in antiquity, someone had the capability and technology needed for quarrying, cutting, and shaping colossal stone blocks in the quarry—then lifting the stone blocks up and carrying them to the construction site, and there not just let go and drop the stone block, but place them precisely in the designated course.*[2]

Ancient Aliens chose to address this issue, too, of the massive stones at Baalbek: "Archaeological surveys have revealed that the enormous stone foundation that lies at the base of the site dates back tens of thousands of years, but even more significant to ancient astronaut theorists is their belief that the colossal stone platform may once have served as a landing pad for space travelers."[3]

A Mars-Earth link

The above *does* amount to just a theory, as were all of Sitchin's thoughts and conclusions in relation to the Anunnaki, White Powder Gold, and near-immortality. That's one part of the story that suggests aliens were the masters of such enigmas as the Baalbek stones and were also.

responsible for raising those goliath-like stones at Stonehenge. Now, though, we're going to take a trip not to another part of the Earth, but to an entirely other world: Mars. The Red Planet.

The planet Mars: a connection with us? (NASA)

The fact is that there is no need, at all, for any theoretical aliens to visit our world from Nibiru, or even from a planet out of our solar system. There are enough flags that demonstrate Mars was a world not at all too dissimilar to ours. Indeed, it's on the Red Planet that we see more than a few enigmas that suggest there is a link between (a) baffling and ancient structures on Earth, and (b) even earlier extraterrestrial similarities on Mars. In some cases we can make a fascinating argument that there are similarities between some of the structures on Mars and others on our world. That *could* mean that we really *were* visited by aliens thousands of years ago. Or, perhaps *even more* intriguing, is the possibility that an extremely ancient human civilization achieved space travel, left their marks on the surface of Mars, and then returned back to the Earth, perhaps hundreds of thousands of years ago.

Okay, it's just a theory that many are happy to ignore, or haven't even considered. The "human angle" however, is a far more plausible theory than one that has aliens from a world—Nibiru—that no one can find. *Anywhere.* But, for now, let's keep looking at the alien theory to explain all of this. At the very least, we should give it some space for debate. Let's do so in relation to what has become known, famously and infamously, as the "Face on Mars."

This mega-scale creation of stone can be found in an area of Mars known as Cydonia. That it appears to resemble a human face and also seems to be surrounded by what look like damaged, massive pyramids only adds to the inflammatory-filled theory that, millennia ago, there was a Mars-Earth link.

Scientific Analysis Would Have to Await Independent Researchers

Mac Tonnies, the author of *After the Martian Apocalypse: Extraterrestrial Artifacts and the Case for Mars Exploration,* spent a great deal of time addressing the matter of that huge face. He told me of his thoughts and his observations; all of them well-thought out but also controversial:

> *When I realized that there was an actual scientific inquiry regarding the Face and associated formations, I realized that this was a potential chance to lift SETI [the Search for Extraterrestrial Intelligence] from the theoretical arena; it's within our ability to visit Mars in person. This was incredibly exciting, and it inspired an interest in Mars itself—its geological history, climate, et cetera. I have a BA in Creative Writing. So, of course, there are those who will happily disregard my book because I'm not "qualified." I suppose my question is "Who*

> *is qualified to address potential extraterrestrial artifacts?" Certainly not NASA's Jet Propulsion Laboratory, whose Mars exploration timetable is entirely geology-driven.*[4]

Tonnies added:

> *NASA itself discovered the Face, on July 25, 1976, and even showed it at a press conference, after it had been photographed by NASA's Viking mission probe. Of course, it was written off as a curiosity. Scientific analysis would have to await independent researchers. The first two objects to attract attention were the Face and what has become known as the "D&M Pyramid." Both of them were unearthed by digital imaging specialists Vincent DiPietro and Gregory Molenaar* [at the time, engineers at NASA's Goddard Space Center at Greenbelt, Maryland]. *Shortly after, Face researcher Richard Hoagland pointed out a collection of features near the Face which he termed the "City." The "Fort," too.*[5]

Also from Tonnies to me: "The Fort looks weathered, defeated. Its eastern side is riddled with small, shallow craters that terminate as abruptly as the holes left from a burst of machine gun-fire. Indeed, it's easy to imagine that you're examining the sterile ruins of some unimaginable conflict."[6]

Indeed, that is what it looks like.

The Beautiful One Has Come

Moving on from the Face on Mars, there's a *second* Face on Mars. They are known as the "Crowned Face" and the "Meridian Face." I have come to know—and to my surprise—that very few people are aware of the existence of these *other* huge, stone faces and their importance to this overall story. The leading figure in this particular story is Greg Orme. He is the author of an excellent book, *Why We Must Go to Mars.* He says of his thoughts and his ideas as to what happened to Mars and the civilization that once lived there:

> *The Meridian Face has some similar features to the Cydonia and Crowned Faces such as having a crown. It is similar to the Nefertiti formation in that it seems to be made of dark soil, dunes in this case. Some dark dune fields can migrate large distances in Meridiani Planum, others remain confined in larger craters perhaps by shielding them from the wind. This can allow for the formation to be very old and remain intact. The similarity between the Crowned Face in the King's Valley Libya Montes and the Meridiani Face was originally shown with an [overlay].*[7]

As for the "Nefertiti formation" that Orme mentions, the *Ancient History Encyclopedia* says Nefertiti

> *was the wife of the pharaoh Akhenaten of the 18th Dynasty of Egypt. Her name means, "the beautiful one has come" and, because of the world-famous bust created by the sculptor Thutmose (discovered in 1912 CE), she is the most recognizable queen of ancient Egypt. She grew up in the royal palace at Thebes, probably the daughter of the vizier to Amenhotep III, a man named Ay, and was engaged to his son, Amenhotep IV, around the age of eleven. There is evidence to suggest that she was an adherent of the cult of Aten, a sun deity, at an early age and that she may have influenced Amenhotep IV's later decision to abandon the worship of the gods of Egypt in favor of a monotheism centered on Aten. After he changed his name to Akhenaten and assumed the throne of Egypt, Nefertiti ruled with him until his death after which she disappears from the historical record.*[8]

That there appears to be a huge image of Nefertiti on Mars is nothing short of incredible.

Mysterious Builders

How about a Stonehenge on Mars? A mind-blowing story surfaced out of the UK's *Daily Express* newspaper on September 24, 2015. A man who chose to use the alias of Mr. Enigma posted to YouTube a picture of what looks eerily, and incredibly, like an overhead photo of Stonehenge. Mr. Enigma told the media that he had found

> *a perfectly circular platform with a strange cluster of stones emerging from it.*
>
> *I know the formation is not an exact match, nor am I saying it is, indeed, a Stonehenge set up. I am just saying there is something strange about this area and it looks very much like the mysterious ancient stone circle of Stonehenge. Could the builders of Stonehenge have visited Mars and did they build the same thing there? Or did we have visitors who taught us how to build these things and do the same for long-lost beings on Mars as well? Or is this just another face on Mars illusion?*[9]

A Sphinx-Style Creation on Mars

Ufologist Nigel Watson chimed in, too. He said of this particular Martian controversy:

> *Pyramid structures have been regularly spotted on Mars that have been linked to the ancient pyramids of Earth, now with Marshenge we have a link with our earlier prehistoric structures. The scorn of the skeptics has failed to erode such theories and with the discovery of Marshenge, it builds an even stronger case to suggest astronauts*

in the past visited our solar system and had an enduring impact on human history.[10]

Last of all on this particular issue, there's the matter of Mars's Sphinx. That's right: There *does* appear to be a Sphinx-style creation on Mars. The photograph is not a fake or a bit of fun; not in the slightest. It's the work of NASA's Jet Propulsion Laboratory (JPL). Although the space agency is content to relegate the image to the world of pareidolia—"seeing faces in clouds," in essence—not everyone is quite sure that's all the photo shows. JPL staff state the following of the image:

The "Twin Peaks" are modest-size hills to the southwest of the Mars Pathfinder landing site. They were discovered on the first panoramas taken by the IMP camera on the 4th of July, 1997, and subsequently identified in Viking Orbiter images taken over 20 years ago. The peaks are approximately 30-35 meters (-100 feet) tall. North Twin is approximately 860 meters (2800 feet) from the lander, and South Twin is about a kilometer away (3300 feet). The scene includes bouldery ridges and swales or 'hummocks' of flood debris that range from a few tens of meters away from the lander to the distance of the South Twin Peak.[11]

A Brief but Catastrophic Flood

Mac Tonnies took a great deal of interest in all of this:

The region is not without its mysteries. In the Pathfinder's landing charts is a third peak simply labeled "North Knob." And while it is certainly knob-like in shape, its perimeter is inscribed by a perfect square. The effect is obvious and uncanny and doesn't appear around the smaller Twin Peaks. North Knob, like the Twin Peaks, is situated on an extinct floodplain, so its square perimeter is doubly surprising; flood features on Mars are typical

> *elliptical or teardrop-shaped, and one can easily tell the direction of the vanished flow by their alignment. The square perimeter may be due to a tough subterranean foundation; the irregular mass on top may be the remains of a structure of some kind that collapsed in the flood, scattering pieces of itself across the plain.*[12]

Mike Bara, whose books include *Ancient Aliens on Mars* and *Dark Mission,* said:

> *Besides all the "near field" Pathfinder anomalies,* [geologist Ron] *Nicks and* [Richard] *Hoagland began studying "super resolution" images of the two distant (almost one kilometer) "mountains" imaged on the horizon of the Landing Site: the celebrated "Twin Peaks." Nicks, in particular, soon realized that these features showed definite signs of engineering, as opposed to natural erosional processes. And although he recognized that the area (as NASA had previously advertised) obviously had been subjected to "a brief but catastrophic flood," he could not explain some of the strikingly geometric features he was seeing on the "Peaks" as "just geology."*[13]

What If They Were a Highly Evolved Body of Extremely Ancient Humans?

Mars, we can suggest, is home to a classic stone circle, pyramids, a "crowned face," two more sculptured heads, a rendition of Nefertiti, and a Sphinx. If just only *one* such anomaly was found by NASA's cameras over the decades to be evidence of some form of intelligence—modern or ancient—then, perhaps, we would be right in dismissing everything as a trick of the light, or of pareidolia. But the truth is this: At some point in our mysterious past, Earth and Mars shared something—something incredible. It was nothing less than the secrets of antigravity, levitation, and the ability to move gigantic stones through the skies. Those secrets, it seems, are now long gone.

Going back to that intriguing theory of an ancient human civilization that may have made it to Mars, possibly tens of thousands of years ago, what if those who created the Earth's incredible structures on both Mars and Earth were not aliens after all? What if they were a highly evolved body of *extremely ancient humans*—the Levitators, one could say—who, incredibly, made it to Mars and even established a civilization on the Red Planet? Mac Tonnies, at one point, pondered on that admittedly outrageous, and mind-blowing, scenario in relation to his research into the Face on Mars and the many and varied other Martian mysteries. Take note of Tonnies's words; they make for brief but fascinating reading.

A Vanished Terrestrial Civilization that Had Achieved Spaceflight

Said Tonnies in an interview with me on July 7, 2009:

> *There's a superficial similarity between some of the alleged pyramids in the vicinity of the Face and the better-known ones here on Earth. This has become the stuff of endless arcane theorizing, and I agree with esoteric researchers that some sort of link between intelligence on Mars and Earth deserves to be taken seriously. That's the disquieting aspect of the whole inquiry; it suggests that the human race has something to do with Mars, that our history is woefully incomplete, that our understanding of biology and evolution might be in store for a violent upheaval.*

Tonnies didn't address this angle too closely, but he did add these words before his death. In the same interview, Tonnies said:

> *In retrospect, I regret not spending more time in* After the Martian Apocalypse *addressing the possibility that the Face was built by a vanished terrestrial civilization that had achieved spaceflight.*

> *That was a tough notion to swallow, even as speculation, as it raises as many questions as it answers.*

Yes, it is hard to swallow, but it shouldn't stop us from tackling that very scenario in the next chapter.

CHAPTER 26

THE CASE FOR AN EARLY, UNKNOWN HUMAN CIVILIZATION

It's intriguing to note that the worlds of antigravity and levitation were discussed, to a significant degree, in the 1950s. Yet there was actually a minimum of talk by that era's Ufologists of an extraterrestrial component behind the construction and movement of all those massive stones. Rather, most of the talk was of ancient, terrestrial civilizations, long-gone humans, and secret (and lost) technology that had been created by the likes of the mythical Atlanteans. We begin with Morris Jessup, the man who dug deep into the world of antigravity and the Philadelphia Experiment, and who died in his car, by carbon-monoxide poisoning, in a Florida park in 1959. As you'll recall, although Jessup was a full-on believer in the UFO phenomenon, he believed the Levitators were very likely from *our* world and *not* extraterrestrials.

On March 27, 1953, Jessup wrote an illuminating letter to his good friend Joseph Manson Valentine (a man who had a deep interest in Atlantean lore and legend), on the *very* theme of this book. Valentine was one of the last people to speak with Jessup. In fact, the two chatted the very night before Jessup's death. In that eye-opening letter, Jessup said something incredible to his friend.

"Flying Saucers Used Some Means of Reacting with the Gravitational Field"

Valentine was becoming more and more impressed—and swayed, too—by Jessup's "ancient humans" theory, and also by Jessup's words and his theories concerning ancient levitation. The outcome was this: Valentine encouraged Jessup to insert a section on antigravity, in ancient times, in the book that Jessup was then working on. Jessup did so in the pages of his to-be-published 1955 book, *The Case for the UFO.*

It's important to note that when we carefully dissect the words of Morris K. Jessup, we find something very intriguing. And somewhat highly surprising, too. Time and time again, Jessup made it very clear in his letters to Valentine (and, later, in his books, *The Case for the UFO* and *The Expanding Case for the UFO*) that he (Jessup) felt antigravity was the key to the mystery of how, in times long past, it all came to the raising of massive stones. Yet, there's something that's *almost* completely missing in much of Jessup's views on antigravity: aliens, extraterrestrials, and beings from other worlds. In fact, Jessup barely touches on such things.

"It Appears that the Original Massive Structures Were Erected by the Earliest Colonists in the Pre-Inca Era"

Jessup said to Valentine:

> *As one reaches back into ancient history and studies the ruins or traces of ancient structures, one discovers that the workmanship, in stone walls, deteriorated steadily with successive buildings. The ramparts, if such they be, of Sacsahuaman in ancient Peru and other structures, were undertakings comparable to our great dams. Handling those vast blocks of stone, so accurately fitted, demanded vast controlled power. Yet those remains seem to have been the earliest. From that prodigious perfection the construction appears*

to have degenerated. In the latter efforts smaller and smaller stones were used and workmanship lost its skill. A few buildings were marvelously well-built while later ones were crude. Eventually all the work declined to a low level.[1]

Jessup continued to Valentine:

True, there are some fairly large stones in the structures of downtown Cuzco, such as the Temple of the Sun, but the largest weigh only a ton or two. True, also, there is superb workmanship in the Temple of the Sun and the House of the Virgins, where stones were almost perfectly squared before being worked into walls; but other walls built on the same principles showed progressively poorer skill, the later stones being scarcely more than large pebbles. One wonders how they stand up. Many are built as structures on the older more solid bases. From this it appears that the original massive structures were erected by the earliest colonists in the pre-Inca era.

"Hundreds of Thousands of Tons"

Jessup had even more to say to Valentine about the "ancient human" issue, but hardly anything about alien-themed aspects of the puzzle:

They brought with them expert knowledge of stonework and a special power or levitating machinery. At least, this is our postulate. Then they must have lost the source of the power and their descendants used hand labor, which couldn't manipulate really big stones. The Black Pagoda of India [the Konark Sun Temple, to be correct, which we'll come to later], *believed by some to have been built seven centuries ago is 228 feet high. Its roof, twenty-five feet thick, is a single stone slab weighing 2,000 tons. A sizable chunk of rock to put on top of a 228-foot building which looks like a silo.*

Historians conjecture that the building was buried in sand, providing a ramp several miles long, up which the roof block was dragged. That is a possibility, of course, and the only method conceivable by any mechanical principle known to present-day engineers. But there might be another possibility. The structure might have been raised under the huge stone, pushing it up a tier at a time, with jacks of some sort. In some ways this looks simpler than building a ramp with hundreds of thousands of tons of sand to bring up and cart away.

Levitation Is the Lesser Quandary

In yet another letter, Jessup shared his then-latest thoughts with Valentine:

I do not think modern engineers could post it on top of such a building without unlimited labor funds to build the ramp. Even then I do not see how they could possibly get sufficient purchase on this thing to move it at all, or attach sufficient harness to apply the necessary force. This incredible stone is larger than those at Baalbek. Levitating it into place seems as credible to me as any known mechanism including a sand ramp.

Maybe this would require that it be more than 700 years old, since levitation would almost certainly have been recorded if employed as recently as 700 years ago. If you were trying to move this thing up a ramp, no matter how gradual, how could you anchor your tackle in the loose sand to get a pull? How would you place enough men around it to push or pull; and how would they secure a toe-hold? Levitation is the lesser quandary.

Jessup was still not done, however. The man was on a definitive roll. As Jessup's further letters to Valentine demonstrated, the issue of levitation continued to fascinate Jessup—some might justifiably say to an obsessive degree—and he continued to pursue this particular theory for the moving of massive stones, but there was scant talk about creatures from other planets. In fact, to put it bluntly, there was *no* such talk. *Yes*, Jessup *was* heavily into the UFO phenomenon, but he was *not* persuaded that those who were flying UFOs in the 1950s were the very same beings that were responsible for those giant stone structures at Baalbek and Egypt. And elsewhere.

"Tenuous Links between Our Immature Revival and the Parent Past"

Several months later, Jessup expanded further to Valentine, who was also quite open to the human angle of the levitation mystery. As a result of his various expeditions and treks, Jessup was seeing evidence of antigravity technology just about everywhere, and he just might have been correct in his ongoing letters to Valentine. Another letter from Jessup, to his good friend, went as follows:

> *All of the centers of civilization and cultural renaissance recognized by present-day anthropologists—India, Peru, Yucatan, Egypt, Babylonia, Greece, China, Rome, England and others—are but the reviving remnants of an empire and civilization which colonized the world a hundred thousand years ago." That was a sensational statement to make, but it wasn't over:*
>
> *They are all parts, or nuclei, in one great renaissance which has been taking place for, roughly, six to ten thousand years. In it are some traces of the archaic, original master culture, and perhaps through India, Tibet, Egypt and Middle America, there are some tenuous links between our immature revival and the parent past. These traces are mostly in the form of stone works, and some glyphs, of a singular nature, with a very few written records existing mostly in the Orient, and particularly in southern Asia.*[5]

Once again: No aliens—or, at the very least, only a few—were in sight for Jessup when it came to the construction of all of those ancient buildings, structures, and stone circles.

"We Find Evidence of Stone Blocks of Unbelievable Weight Being Quarried"

At this point, also in early 1953, one can see that Jessup went on something of a rant, albeit a rant that made his point carefully and to the point:

> *All of this is anathema to conventional science, archaeology and anthropology especially, for organized science has set up a pattern which covers human growth in broad general terms and has accepted some rigidly restricting tenets which limit original thinking and shut out the obvious. While these general assumptions of science are largely proven by observation and deduction, they are only proven to a point. Beyond that point there are the "erratics": little annoying things, events, or artifacts, which stubbornly refuse to fit into the pattern, and which are sturdily disregarded in the interest of maintaining a working hypothesis acceptable to science in its current state of thinking.*

It was after this particular letter reached Valentine that Jessup went global with his theories, so to speak. He said to Valentine, on May 3, 1954:

> *In many areas we find evidence of stone blocks of unbelievable weight being quarried, more or less casually moved considerable distances, then lifted into place. This common factor connects pre-Inca Peru with Easter Island in a startling and undeniable way, and seems to tie in the Middle East, the Orient, Africa, and maybe Polynesia.*

Many investigators and thinkers have proposed methods for moving these quarried and dressed blocks. All of the proposals are based on application of such simple present-day engineering equipment as block-and-tackle or sand ramps. The great pyramids, consisting of hundreds of thousands of slaves toiling up long ramps of sand to bring these gigantic masses from the Nile. Flotation has been considered. No suggestions have been made which really fit all cases, and some of the submissions are so cumbersome and inadequate as to seem ridiculous.

Lost

At one point, also in 1954, Jessup alluded to an alien connection, but it was something that he hardly championed. Really, it wasn't much more than your average aside:

I have used the word "levitation" as a substitute for power or force. I have suggested that flying saucers used some means of reacting with the gravitational field. In this way they could apply accelerations or lifting forces to all particles of a body, inside and outside, simultaneously, and not through external force applied by pressure, or harness, to the surface only. I believe that this same, or a similar force was used to move stones in very ancient times. I believe the source of this lifting or levitating power was lost suddenly.

As those words directly above make it very clear, Jessup was extremely careful as to how he structured his statement. And to how he presented his thoughts.

Getting to the End of the Theories

What might have caused that sudden loss of antigravity technology that Jessup referred to in the statement above? Jessup, certainly, offered no answer to that question. He didn't even provide a theory. In fact, Jessup's words imply—to me, at least—that, though aliens *may* very well have been involved in the construction of some of our huge, old structures, it was ancient humans who had somehow *acquired* incredible stone-raising technology. Who mastered the science behind levitation, and who were responsible for the incredible movement of huge stones around the world, and millennia ago? Aside from just a sentence or two, levitation and aliens—together—barely crossed paths in the strange world of Morris Jessup. For the man, himself, the human equation was *always* the most important one.

It was these many detailed letters between Jessup and Valentine in 1953 that the former saved from destruction and eventually chose to place their content into 1955's *The Case for the UFO*. Had Jessup decided to destroy all of that priceless correspondence back in 1953, matters would, possibly, be very different.

CONCLUSIONS

Now, it's time for us to come to a conclusion on the mysterious matter of ancient, acoustic levitation. As we've seen, levitation and antigravity have played significant roles in the creation—and the movement—of absolutely massive stones that can be seen all across the world. Certainly, one of the most important things about the story you have just read is that it's not localized. And it never has been. A careful study of the history of the phenomenon of acoustic levitation shows that it's a global mystery, and a mystery that can be pierced—if, that is, we know exactly where to look. It's important to note that stories of stones floating, or perhaps even flying, through the skies can be found all across the planet. Those stories are attributed to mysterious people who, so far as we know, had no connections to each other. Yet, based on all of that incredible architecture, connections are *exactly* what there seemed to have been.

The most fascinating, and baffling, aspect of all this is that while stories of stones careering through the skies abound, we lack the one key thing that would nail all of the stories to the wall: technology. That's right: Aside from the weird tales coming out of Egypt—of the use of papyrus and iron rods to elevate multi-ton stones (clearly, just a legend distorted wildly out of an ancient reality)—there's nothing to demonstrate that amazing, high-tech machines were *ever* used to raise massive stones. Not a solitary battered or dented "antigravity machine" has ever been found in the hot, desert land at Giza. No telltale rusted "levitation device" has ever been dug up by a team of archaeologists at Uxmal. The very same thing goes for Stonehenge. Has anyone ever stumbled on an "acoustic gun" to bring down cities? Not a one. In short, what has been found in all of this is . . . *nothing.* This very much echoes the strange story of Edward Leedskalnin, particularly in the 1940s and into the very early 1950s. He achieved an *enormous* amount while working with massive amounts of stone, but—both eerily and strangely—he did so in a way that no one else could figure out.

This curious picture suggests several things: (a) that the total lack of any meaningful ancient technology is due to the stories of the world's floating stones being nothing but folklore and mythology that, over time, got spread across the world by word of mouth; (b) that the stories of antigravity in the past are overwhelmingly true, but that the raising of rocks in eye-popping fashion thousands of years ago *didn't require the use of technology as we understand it*; (c) that we don't give our ancestors the respect we should when it comes to their abilities to move massive weights across difficult landscapes; and, now, (d): that, eventually, any actual technology that may have existed—to raise the stones—was either deliberately hidden away or lost to the fog of time.

There is something else, too: It's very obvious that the ancient people of those long-gone eras were completely driven by sound. It dictated their lives. Whether the work of humans or of extraterrestrial Levitators, sound completely ruled—and across much of the Earth. There was acoustic levitation. And there was infrasound. In ancient times, sound could bring down a city. It had the awesome power to open the Red Sea. Sound was vital to medical treatment back then. Sound played a role in religious ceremonies. Meditation, too. Blocks of stone that weighed more than 1,000 tons were part and parcel of their lives. Imagine a world—and a science—so different to ours. In many respects, it's nearly impossible to conceive what such a world, so controlled by sound, would have really been like.

Finally, and by now, you can't fail to see something important: Throughout these pages, I have deliberately kept my views on who the Levitators really were to a minimum. It's all very simple: I have no proof that long ago there were ancient Martians who visited our world. Unlike Zecharia Sitchin, who went totally over the top when it came to blurring fact and fiction, I have not flown the flag of Nibiru. I've no proof that the Anunnaki were all behind this—or even if they really lived. Rather, I've chosen to comment on what I've found and, in the process, sought to get as many answers as possible. The purpose for all of this restraint? It's very simple: In this story, we actually know very little. All we know is that thousands of years ago there were races who could achieve incredible things of a kind that we simply cannot.

There were times when they were here. There were times when they left behind them amazing creations that were so incredible, huge, and intricate, they will never be forgotten. They were mysterious and enigmatic. They were the Levitators. And that's what I know.

CHAPTER NOTES

Introduction

1. Jones and Flaxman, *The Resonance Key.*
2. Wilson, "How Acoustic."

Chapter 1

1. Kolosimo, *Timeless Earth.*
2. Cathie, Bruce L., and Peter N. Temm. *UFOs and Anti-Gravity.*
3. Cathie, Bruce
4. Cathie, *The Harmonic Conquest of Space.*
5. Ibid.
6. Mathers, "The Key."
7. Wedeck, A Treasury.
8. Ibid.

Chapter 2

1. Boeche.
2. Boeche. January 22, 2007 interview.

Chapter 3

1. "Fly Agaric."
2. Ibid.
3. Hein, "What Is?"
4. Puharich, *The Sacred Mushroom.*

Chapter 4

1. "Robert M. Schoch," and Ivimy, *The Sphinx.*
2. Owen, "Ancient Egypt."
3. Prakash, "Egypt."
4. "Robert M. Schoch."
5. "Robert M. Schoch."
6. Ibid.
7. Ibid.
8. Ibid.
9. Hemeda and Sonbol, "Sustainability."
10. Konney, Kara. "How Did Egyptians Build the Pyramids?" *www.history.com*

Chapter 5

1. "Easter Island." *history.com.*
2. Ibid.
3. Clark, "Secrets."
4. Schwartz and Bair, *The Mysteries of Easter Island.*
5. Clark, "Secrets."
6. "Easter Island Moai."
7. Hirst, "How the Moai."
8. "Easter Island Moai."
9. Sellier, *Mysteries of the Ancient World.*
10. Ibid.
11. Ibid.
12. Ibid.
13. Ibid.
14. "History of Easter Island."
15. Petricevic, "Unwritten."
16. "Göbekli Tepe."

Chapter 6

1. Mark, "Baalbek."
2. Ibid.

3. "Baalbek."
4. Ibid.
5. Sitchin, Zecharia. "In the News: Baalbek. Welcome to the 'Landing Place.'" *sitchin.com/landplace.htm*
6. "Anunnaki Structures."
7. Gardner, *Genesis of the Grail Kings.*
8. Batuman, "The Myth."
9. Cline, "Roman Heliopolis."
10. David, "Baalbek Lebanon."
11. "Ba'albek."
12. Ibid.
13. Ibid.
14. Ibid.
15. Ibid.
16. Ibid.
17. Ibid.
18. Ibid.
19. "Baalbek Trilithon."

Chapter 7

1. McDermott, "The Spectacular."
2. Ibid.

Chapter 8

1. Bord and Bord, *The Secret Country.*
2. Ibid.
3. "History of Kit."
4. Ibid.
5. Volpe, "Highgate."
6. Trevelyan, *Folk-Lore and Folk-Stories of Wales.*
7. Ibid.
8. Lewis, "The Cheesewring."
9. Ibid.

Chapter 9

1. "Overview and History."
2. Ibid.
3. Ibid.
4. "The Story."
5. Robins, *Circles of Silence.*
6. "Featured Author."
7. Ibid.

Chapter 10

1. German, "Stonehenge."
2. "Stonehenge." (English Heritage.)
3. "Stonehenge." (History in Numbers.)
4. Ibid.
5. Plackett, "Why?"
6. Ibid.
7. "Stonehenge." (English Heritage.)
8. "Preseli Hills."
9. Pearson et al., "The Original."
10. De Camp, *Great Cities of the Ancient World.*
11. "Stonehenge Bluestones."
12. Satter, "UK Experts."
13. Stetka, "Sound Waves."
14. "Archaeoacoustics." *www.bibliotecapleyades.net.*
15. Gallagher, James, "Ultrasound May Heal Chronic Wounds, Suggest Study." *www.bbc.com.*
16. Gallagher, "Ultrasound."
17. Ibid.
18. Ibid.

Chapter 11

1. "Red Sea."
2. "What Is the Meaning?"

3. Nuwer, "The Science."
4. "Acoustic Levitation." (BBC.)
5. Redfern, "Commanding."
6. Ibid.
7. Kenyon, "Jericho."
8. "Battle of Jericho."
9. Jarus, "What Is?"
10. Ibid.
11. Isaacs, *Talking with God.*
12. Ibid.
13. Tesla, "The Wonder." *teslauniverse.com*
14. "The Wonder World."
15. Pilkington, "Ancient Electricity."

Chapter 12

1. Hitching, *Earth Magic.*
2. All quotations in this paragraph are from Hitching, *Earth Magic.*
3. De Camp, *Great Cities of the Ancient World.*
4. Ibid.
5. Dritsas, "Music Therapy."
6. "Earliest References."
7. Noorbergen, *Secrets of the Lost Races.*
8. Ibid.
9. Ibid.
10. Lowe, "Suspending."
11. Charroux, *One Hundred Thousand Years of Man's Unknown History.*
12. Ibid.
13. Berlitz, *Mysteries from Forgotten Worlds.*
14. Tomas, *We Are Not the First.*
15. Lewis, *Footprints on the Sands of Time.*
16. Ibid.
17. Ibid.
18. Ibid.

19. Fix, *Pyramid Odyssey.*
20. Wagner, "Levitation."
21. Mooney, *Colony Earth.*

Chapter 13

1. "Daniel Dunglas Home" and "Home, D.D."
2. "Daniel Dunglas Home," "Home, D.D.," and the author's personal research.
3. Ibid.
4. *The Quarterly Journal.*
5. "Flying In and Out of Windows." *strangehistory.net*
6. "William Stainton."
7. Summers, *The History of Witchcraft.*
8. Hocking. *The Purpose of Life, Why Are We Here. books.google.com*
9. Ibid.
10. "Daniel Dunglas Home."

Chapter 14

1. Redfern, *The Slenderman Mysteries.*
2. Guy, "Alexandra."
3. Tomas, We Are.
4. Redfern, *The Slenderman Mysteries.*
5. Tomas, *We Are Not the First.*
6. Cohen, *The Ancient Visitors.*
7. Smythies in Childress, *Technology of the Gods. books.google.com.*
8. "The Flying."
9. Collins, *Gods of Eden.*
10. Joffe, "Tripping."
11. Collins, *Gods of Eden.*

Chapter 15

1. "Coral Castle."
2. Castrillo, Karina, "The Story Behind Ed Leedskalnin's Carol Castle in Florida." Champion, Bonfils, and Brady, "Coral Castle." *Roadside America.* "Coral Castle." Random Times, "Edward Leedskalnin." Winston, Steve. "Homestead's Coral Castle.".
3. Winston, "Homestead's."
4. "Edward Leedskalnin."
5. Ibid.
6. Ibid.
7. Ibid.
8. Castrillo, "The Story."
9. Champion, Bonfils, and Brady, "Coral."
10. Radford, "Mystery."
11. The story in this section comes from Miller, "Coral." There are many other articles that tell the story of the hospital as well.
12. Sethi, "Coral."

Chapter 16

1. Gorvett, "This Temple."
2. All of the quotations throughout this section between Jessup and Manson Valentine, a friend and researcher of Jessup, are from letters, mostly undated. For years the letters were held by a ufologist named Robert C. Davis of Dallas, Texas. Davis had them until a few years ago, when he died. Jessup went on to use all the Valentine letters in his (Jessup's) books *The Case for the UFO* and *The Expanding Case for the UFO.*
3. Moore and Berlitz, *The Philadelphia Experiment,* and Genzingler, *The Jessup Dimension.*
4. Ibid.
5. Cleaver, "Anti-Gravity."
6. Ibid.
7. Ibid.
8. "Philadelphia Experiment."

Chapter 17

1. Cathie and Temm. *UFOs and Anti-Gravity.*
2. Pilkington, "Flight of Fancy." *theguardian.com*
3. Cathie, *The Harmonic Conquest of Space.*
4. "About DIA."
5. Good, *Above Top Secret.*
6. The source is a document declassified by the Defense Intelligence Agency under the Freedom of Information Act that contains correspondence with the DIA and with Bruce Cathie.
7. Ibid.
8. Ibid.
9. Ibid.
10. Ibid.
11. "USNS Eltanin."
12. See endnote #6 in this chapter.

Chapter 18

1. Borneman,"What Are?"
2. Michell, "Watkins's."
3. Ibid.

Chapter 19

1. Bellamy and Walker, "Secret."
2. "MoD Boscombe Down."
3. Noyes made the statement in the very popular Barge Inn pub in Wiltshire, England. The Barge Inn is where many Crop Circle researchers hang out. And Noyes did a lot of research into Crop Circles.

Chapter 20

1. Tingley, "Scientists."
2. Andrade, Bernassau, and Adamowski, "Acoustic."
3. Kennedy, "Scientists."
4. Dajose, "Levitating."

5. Ibid.
6. Seaburn. *mysteriousuniverse.org*
7. Marzo, Caleap, and Drinkwater, "Acoustic."
8. "Acoustic Levitation." (themanews.)
9. Andrade, Bernassau, and Adamowski, "Acoustic."
10. Ibid.
11. Ibid.
12. Ibid.
13. University of Bristol.
14. "Acoustic Levitation." (BBC.)

Chapter 21

1. Most of the story comes from the fact that I know the story myself, from my own research over decades. Interested readers can also look at Screeton, Quest.
2. Cartwright, "Ancient."
3. Ibid.
4. Davis, "Sleep."
5. Boothman, "The Hexham."
6. Ibid.
7. Robins, *Circles of Silence.*
8. Ibid.
9. Lucia. "How Does?"
10. "The Stone Tape."
11. "Nine Ladies."
12. "Stanton Moor."
13. Downes, *Monster Hunter.*
14. Harpur, *Mystery Big Cats.*
15. Marrs, *PSI Spies.*
16. Redfern, *Nessie!*
17. "Clava Cairns."
18. Castelow, "The Battle."
19. Robinson, *The Monsters of Loch Ness.*
20. Ibid.

Chapter 22

1. "Infrasound."
2. "Why Do?"
3. Hamer, "Tigers."
4. Muggenthaler, "Low."
5. "Sounds."
6. Associated Press, "Infrasound."
7. Ibid.
8. Ibid.
9. "Bigfoot."
10. Ibid.
11. MissSquatcher, "The Infrasonic."
12. Godfrey, *Hunting the American Werewolf.*
13. Horowitz, "Could?"

Chapter 23

1. "All About."
2. "History of Avebury."
3. The source of this story is Malcolm Lees, a retired UK military personnel. The story came from a file that Lees was able to read in the early 1960s. All declassified UK government files are now held at the National Archives not far from London.
4. Ibid.
5. Bord and Bord. *Mysterious Britain.*

Chapter 24

1. Adams and Williams. *Biological Effects of Electromagnetic Radiation (Radiowaves and Microwaves) Eurasian Communist Countries.*
2. Ibid.
3. Ibid.
4. "RAF Greenham."
5. Redfern, "A Strange."
6. Parry, "Peace."

7. Keith, *Mind Control and UFOs.*
8. Ibid.
9. Ibid.
10. Radford, "What Is?"
11. "Taos Hum."
12. "What Is Causing?"
13. "The Strange Case."
14. Ibid.
15. "Raytheon."
16. Liszewski, "Future."
17. Carr, "Understanding."

Chapter 25

1. "White Powder."
2. *www.bibliotecapleyades.net*
3. Hale, *The Ancient Alien Theory.*
4. Tonnies, *After the Martian Apocalypse.*
5. Ibid.
6. Ibid.
7. Orme, "The Meridiani."
8. Mark, "Nefertiti."
9. Austin, "Has Stonehenge?"
10. Gray, "Stonehenge-Style."
11. "Twin Peaks."
12. Tonnies, *After the Martian Apocalypse.*
13. Hoagland and Bara. *Dark Mission.*

Chapter 26

1. All of the quotations throughout this section between Jessup and Manson Valentine, a friend and researcher of Jessup, are from letters, mostly undated. For years the letters were held by a ufologist named Robert C. Davis of Dallas, Texas. Davis had them until a few years ago, when he died. Jessup went on to use all the Valentine letters in his (Jessup's) books *The Case for the UFO* and *The Expanding Case for the UFO.*

BIBLIOGRAPHY

"About DIA." Defense Intelligence Agency website. *dia.mil.*

"Acoustic Levitation: Chemical Reaction Lifted by Sound." BBC News July 16, 2013. *bbc.com.*

"Acoustic Levitation: Floating on a Wave of Sound?" themanews. October 8, 2019. *en.protothema.gr/.*

Adams, Ronald L., and R.A. Williams. *Biological Effects of Electromagnetic Radiation (Radiowaves and Microwaves) Eurasian Communist Countries.* Defense Intelligence Agency. October 10, 1975. *ecfsapi.fcc.gov. www.dia.mil.*

Albarelli, H.P., Jr. *A Terrible Mistake: The Murder of Frank Olson and the CIA's Cold War Experiments.* Walterville, Oregon: Trine Day LLC, 2009.

"All about Avebury." History Press. 2021. *thehistorypress.co.uk.*

Andrade, Marco, A.B., Anne L. Bernassau, and Julio C. Adamowski. "Acoustic Levitation of a Large Sphere." AIP. July 26, 2016. *scitation.org.*

"Anunnaki." Library of Halexandria. February 5, 2009. *halexandria.org.*

"Anunnaki Structures." Sitchin Studies website. *www.sitchinstudies.com/.*

"Archaeoacoustics." Academic. 2021. *en-academic.com.*

Associated Press. "Infrasound Linked to Spooky Effects." NBC News. September 7, 2003. *nbcnews.com.*

Atwater, Harry A., and Ognjen Ilic. "Self-Stabilizing Photonic Levitation and Propulsion of Nanostructured Macroscopic Objects." March 22, 2019. *nature.com.*

"Aubrey Holes." History in Numbers. 2021. *historyinnumbers.com.*

Austin, Jon. "Has Stonehenge Been Found on Mars? Ancient 'Alien' Stone Circle Discovered on Red Planet." Express. September 24, 2015. *express.co.uk.*

"Baalbek." *ancientaliensdebunked.com.*

"Baalbek Trilithon." Atlas Obscura. *atlasobscura.com.*

"Ba'albek." Twain's Geography. *twainsgeography.com.*

"Battle of Jericho." Bible Study Tools. 2021. *biblestudytools.com.*

Batuman, Elif. "The Myth of the Megalith." *New Yorker.* December 18, 2014. *newyorker.com.*

Bayley, Harold. *Archaic England: An Essay in Deciphering Prehistory from Megalith Monuments, Earthworks, Custos, Coins, Place-Names, and Faerie Superstitions.* London: Chapman & Hall, Ltd., 1919.

Bellamy, Christopher, and Timothy Walker. "Secret US Spyplane Crash May Be Kept under Wraps." March 14, 1997. *independent.co.*

Bergier, Jacques. *Mysteries of the Earth: The Hidden World of the Extra-Terrestrials.* London: 1977.

Berlitz, Charles. *Mysteries from Forgotten Worlds.* New York: Dell Publishing Co., Inc., 1973.

"Bigfoot and Infrasound." The Bigfoot Field Journal. *bf-field-journal.blogspot.com.*

Bingham, Hiram. *Lost City of the Incas: The Story of Machu Picchu and its Builders.* Lima, Peru: Librerias A.B.C. S.A., 1975.

Boeche, Dr. Raymond W. website. *rayboeche.academia.edu.*

Boothman, Neil. "The Hexham Heads." Mysterious Britain & Ireland. Updated December 9, 2018. *mysteriousbritain.co.uk.*

Bord, Colin, and Janet Bord. *The Secret Country: A New Focus on the Folklore and Mysterious Stones of Ancient Britain.* New York: Walker & Company, 1976.

Bord, Janet, and Colin Bord. *Mysterious Britain: Ancient Secrets of the United Kingdom and Ireland.* Aylesbury, UK: Paladin Books, 1983.

Borneman, Elizabeth. "What Are Ley Lines?" GeographyRealm. October 24, 2014. *geographyrealm.com.*

"'Bristol Hum': Residents Report Return of Mysterious Noise." Independent. January 19, 2016. *independent.co.uk.*

"Building Stonehenge." English Heritage. 2021. *english-heritage.org.*

Carr, Daphne. "Understanding the LRAD, the 'Sound Cannon' Police Are Using at Protests, and How to Protect Yourself from It." Pitchfork. June 9, 2020. *pitchfork.com.*

Cartwright, Mark. "Ancient Celts." World History. April 2021. *worldhistory.org.*

Castelow, Ellen. "The Battle of Culloden." Historic UK. 2021. *historic-uk.com.*

Castrillo, Karina. "The Story Behind Ed Leedskalnin's Coral Castle in Florida." Culture Trip. March 22, 2018. *theculturetrip.com.*

Cathie, Bruce. *The Energy Grid: Harmonic 695.* Kempton, Illinois: Adventures Unlimited Press, 1997.

———. *The Harmonic Conquest of Space.* Kempton, Illinois: Adventures Unlimited Press, 1998.

Cathie, Bruce L., and Peter N. Temm. *UFOs and Anti-Gravity.* San Francisco, California: Strawberry Hill Press, 1977.

Champion, Sam, Darcy Bonfils, and Jonann Brady. "Coral Castle: Mysterious Monument to Lost Love." ABC News. July 29, 2008. *abcnews.go.com.*

Charroux, Robert. *Legacy of the Gods.* Aylesbury, UK: Sphere Books, Ltd., 1979.

———. *Masters of the World.* London: Sphere Books, Ltd., 1979.

———. *The Mysterious Past.* New York: Berkley Publishing Corporation, 1975.

———. *One Hundred Thousand Years of Man's Unknown History.* New York: Berkley Publishing Corporation, 1971.

"Children of the Stones." Nostalgia. 2021. *nostalgiacentral.com.*

Childress, David Hatcher. *Anti-Gravity and the Unified Field.* Stelle, Illinois: Adventures Unlimited Press, 1992.

Clark, Liesl. "Secrets of Easter Island." Nova. November 2000. *www.pbs.org.*

"Clava Cairns." VisitScotland. 2021. *visitscotland.com.*

Cleaver, A.V. "Anti-Gravity Booming." *Aero Digest.* March 1956.

Cline, Austin. "Roman Heliopolis & Temple Site at Baalbek in Lebanon's Beqaa Valley." Learn Religions. June 25, 2019. *learnreligions.com.*

Cohen, Daniel. *The Ancient Visitors: Have Creatures from Other Planets Ever Landed on Earth?* New York: Doubleday & Company, Inc., 1976.

Colligan, Douglas. *Strange Energies, Hidden Powers.* New York: Scholastic Book Services, 1979.

Collins, Andrew. *Gods of Eden: Egypt's Lost Legacy and the Genesis of Civilization.* Rochester, Vermont: Bear & Company, 2002.

"Coral Castle." 2021. *roadsideamerica.com.*

Dajose, Lorinda. "Levitating Objects with Light." Caltech. March 18, 2019. *caltech.edu.*

"Daniel Dunglas Home." *PSI Encyclopedia.* 2021. *psi-encyclopedia.spr.ac.uk.*

Dash, Mike. *Borderlands: The Ultimate Exploration of the Unknown.* New York: Random House, 2000.

David. "Baalbek Lebanon—Roman Ruins, Temple, & Megalith Stones." World Travel Guy. December 17, 2019. *theworldtravelguy.com.*

David, Jay. *Flying Saucers Have Arrived! 30 Documented Reports.* New York: The World Publishing Company, 1970.

David-Neel. *Magic and Mystery in Tibet.* New York: University Books, 1958.

Davis, Susan. "Sleep Paralysis: Demon in the Bedroom." WebMD. June 21, 2021. *webmd.com.*

De Camp, L. Sprague. *The Ancient Engineers: The Builders of Egypt, Babylon, Greece and India, Who Were They and How Did They Do It?* New York: Ballantine Books, 1974.

———. *Citadels of Mystery: The Past Yields Its Secrets Reluctantly . . . Clues to the Ways of Ancient Knowledge Are Found in the Ruins of Twelve Civilizations.* New York: Ballantine Books, 1973.

———. *Great Cities of the Ancient World.* New York: Doubleday & Company, Inc., 1972.

Devereux, Paul. *Stone Age Soundtracks: The Acoustic Archaeology of Ancient Sites.* London: Vega, 2001. *www.history.com.*

Dingwall, Dr. Eric J., and John Langdon-Davies. *The Unknown—Is It Nearer?* New York: The New American Library, Inc., 1968.

Downes, Jonathan. *Monster Hunter: In Search of Unknown Beasts at Home & Abroad.* Woolsery, UK: CFZ Press, 2004.

Dritsas, Athanasios. "Music Therapy in Ancient Greece." Greece Is, January 9, 2017. *greece-is.com.*

"Earliest References to Music Therapy." The History of Music and Art Therapy website. *musicandarttherapy.umwblogs.org.*

"Easter Island." Britannica. 2021. *britannica.com.*

"Easter Island." History.com. February 28, 2020. *history.com.*

"Easter Island Moai." Khan Academy. 2021. *khanacademy.org.*

"Edward Leedskalnin and the Mysteries of Coral Castle." Random Times. November 12, 2019. *random-times.com.*

Edwards, Frank. *Stranger Than Science.* New York: Lyle Stuart, 1959.

"'El Gigante' and the Stone Moai of Easter Island." Atlas Obscura. 2021. *atlasobscura.com.*

"Eltanin." Columbia. 2021. *columbia.edu.*

Falde, Nathan. "Acoustic Levitation: Floating on a Wave of Sound." Ancient Origins. June 16, 2019. *ancient-origins.net.*

"Featured Author: Paul Devereux." *megalithic.co.uk.*

Fix, William R. *Pyramid Odyssey: Dramatic New Evidence Reveals the Ancient Secret.* Urbanna, Virginia: Mercury Media, Inc., 1984.

"Fly Agaric." U.S. Forest Service. 2021. *www.fs.fed.us.*

"The Flying Mystics of Tibetan Buddhism." Oglethorpe University Museum of Art. January 25, 2004. *oglethorpe.edu.*

"Former Combat Support Building (Building 273), Greenham Common." Historic England. 2020. *historicengland.com.*

Furneaux, Rupert. *The World's Strangest Mysteries.* New York: Ace Books, 1961.

Gallagher, I.J. *The Case of the Ancient Astronauts.* London: Raintree Childrens Books, 1977.

Gallagher, James. "Ultrasound May Heal Chronic Wounds, Suggests Study." BBC News, July 13, 2015. *bbc.com.*

Gardner, Lawrence. *Genesis of the Grail Kings.* Beverly, Mass.: Fair Winds Press, 2002.

Gaunt, Bonnie. *Stonehenge: A Closer Look.* New York: Bell Publishing Company, 1979.

Genzingler, Anna Lykins. *The Jessup Dimension.* Clarksburg, West Virginia: Saucerian Press, 1981.

German, Dr. Senta. "Stonehenge." Khan Academy. *khanacademy.org.*

"Göbekli Tepe." UNESCO World Heritage Centre website. *whc.unesco.org.*

Godfrey, Linda S. *Hunting the American Werewolf: Beast Men in Wisconsin and Beyond.* Madison, Wisconsin: Trail Books, 2006.

Good, Timothy. *Above Top Secret: The Worldwide UFO Cover-Up*. London: Sidgwick & Jackson, 1987.

Goodman, Jeffrey. *Psychic Archaeology: Time Machine to the Past—Breakthrough: A Scientist-Psychic Team Uncovers an Incredible Ancient Civilization.* Tucson, Arizona: Berkley Medallion Books, 1978.

Gorvett, Zaria. "This Temple at Cholula Dwarfs the Great Pyramid at Giza, Yet It Went Unnoticed by Spanish Invaders. Why?" BBC. August 12, 2016. *bbc.com.*

Gray, Richard. "Stonehenge-Style Rocks Spotted on MARS: Bizarre Circular Stone Formation on the Red Planet Resembles the Iconic Pagan Site." *Daily Mail.* Updated September 24, 2015. *dailymail.co.uk.*

"Greg Orme: Biography." *Coast to Coast AM.* 2019.

Guy, David. "Alexandra David Neel." *Dhamma Wheel.* Fall 1995. *tricycle.org.*

Hadingham, Evan. *Circles and Standing Stones: An Illustrated Exploration of the Megalith Mysteries of Early Britain.* New York: Walker and Company, 1975.

Hale, C.R. *The Ancient Alien Theory: Part Three.* Lulu.com. 2018.

Hamer. Mick. "Tigers Use Infrasound to Warn off Rivals." *New Scientist.* May 2, 2003. *newscientist.com.*

Harpur, Merrily. *Mystery Big Cats.* Heart of Albion Press, 2006.

Harrison, Harry, and Leon E. Stover. *Stonehenge.* New York: Manor Books, Inc., 1975.

Haynes, Renee. *The Hidden Springs: An Enquiry into Extra-Sensory Perception.* Little, Brown & Company, 1973.

Hein, Simeon. "What Is Remote Viewing?" Gaia. November 21, 2019. *gaia.com.*

Hemeda, Sayed, and Alghreeb Sonbol. "Sustainability Problems of the Giza Pyramids." *Heritage Journal.* January 30, 2020. *heritagesciencejournal.springeropen.com.*

Hill, Rosemary. *Stonehenge: Wonders of the World.* Cambridge, Massachusetts: Harvard University Press, 2013.

Hirst, Kris K. "How the Moai of Easter Island Were Made and Moved." ThoughtCo. November 24, 2019. *thoughtco.com.*

"History of Avebury Henge and Stone Circles." English Heritage. 2021. *english-heritage.org.uk.*

"History of Easter Island." Easter Island Travel. 2021. *easterisland.travel.com.*

"History of Kit Coty's House and Little Kit's Coty House." English Heritage. *english-heritage.org.uk.*

Hitching, Francis. *Earth Magic: The Astounding Mystery of the Greatest of All Lost Civilizations.* New York: William Morrow and Company, Inc., 1977.

Hoagland, Richard C., and Mike Bara. *Dark Mission: The Secret History of NASA.* Port Townsend, Washington: Feral House, 2009.

Hocking, M.G., *The Purpose of Life, Why Are We Here. books.google.com*

"Home, D.D. (Daniel Dunglas) 1833-1886." WorldCat Identities. *worldcat.org.*

Horowitz, Seth S. "Could a Sonic Weapon Make Your Head Explode?" *Popular Science.* November 20, 2012. *popsci.com.*

Hutin, Serge. *Alien Races and Fantastic Civilizations.* New York: Berkley Medallion Books, 1970.

"Infrasound: The Noise You Feel." Audio Visual. October 31, 2017. *avbend.com.*

Isaacs, Roger D. *Talking with God: The Radioactive Ark of the Testimony. Communication through It. Protection from It.* Chicago, Illinois: Sacred Closet Books, 2010.

Ivimy, John. *The Sphinx and the Megaliths.* New York: Harper & Row Publishers, 1975.

Jairazbhoy, R. A. *Ancient Egyptian Survivals in the Pacific.* London Jarus: Karnak House, 1990. Jausn. "Ancient Egypt: A Brief History." Live Science. July 28, 2016. *www.livescience.com.*

Jarus, Owen. "What Is the Ark of the Covenant?" Live Science. March 6, 2019. *livescience.com.*

Jessup, Morris K. *The Case for the UFO: Unidentified Flying Objects.* New York: The Citadel Press, 1955.

———. *The Expanding Case for the UFO.* New York: The Citadel Press, 1957.

Joffe, Ben. "Tripping on God Vibrations: Cultural Commodification and Tibetan Singing Bowls." Savage Minds. October 31, 2015. *savageminds.org*

Jones, Marie, and Larry Flaxman. *The Resonance Key: Exploring the Links Between Vibration, Consciousness, and the Zero Point Grid.* Pompton Plains, New Jersey: New Page Books, 2009.

Keith, Jim. *Mind Control and UFOs.* Kempton, Ilinois: Adventures Unlimited Press, 2005.

Kennedy, Sequoyah. "Scientists Say They've Figured Out How to Levitate Objects Using Only Light." Mysterious Universe. March 25, 2019. *mysteriousuniverse.org.*

Kenyon, Kathleen Mary. "Jericho." *Britannica.* 2021. *britannica.com.*

Kolosimo, Peter. *Timeless Earth: What Is the Real Age of Human Civilization? Have We Been Influenced by Creatures from Outer Space?* Secarus, New Jersey: University Books Inc., 1974.

Landsburg, Alan. *In Search of Lost Civilizations.* New York: Bantam Books, 1976.

Lehner, Mark. *The Egyptian Heritage.* Virginia Beach, Virginia: A.R.E. Press, 1974.

Lewis, L. M. *Footprints on the Sands of Time.* New York: New American Library, 1975.

Lewis, Scott Diane. "The Cheesewring on the Bodmin Moor." March 13, 2014. *englishhistoryauthors.blogspot.com/.*

Liszewski, Andrew. "Future Riot Shields Will Suffocate Protestors with Low Frequency Speakers." December 15, 2011. *gizmodo.com.*

Lowe, Dunstan. "Suspending Disbelief: Magnetic and Miraculous Levitation from Antiquity to the Middle Ages." JSTOR website. *jstor.org.*

Lucia. "How Does it Work?: The Stone Tape Theory, Residual Hauntings, and the Deep Influence of Memory and Emotion." March 9, 2020. *theghostinmymachine.com.*

Mark, Joshua J. "Baalbek." World History. September 2, 2009. *worldhistory.org.*

———. "Nefertiti." World History. April 14, 2014. *worldhistory.org.*

Marrs, Jim. *PSI Spies: The True Story of America's Psychic Warfare Program.* Newburyport, Massachusetts: Weiser Books, 2007.

Marzo, Asier, Mihai Caleap, and Bruce W. Drinkwater. "Acoustic Virtual Vortices with Tunable Orbital Angular Momentum for Trapping of Mie Particles." January 22, 2018. *aps.org.*

Mathers, S.L. MacGregor. "The Key of Solomon the King." Kalamazoo Public Library. *kpl.org.*

McDermott, Alicia. "The Spectacular Ancient Maya City of Uxmal." Ancient Origins. August 31, 2018. *ancient-origins.net.*

Meyer, David. "Ark of the Covenant: Lost Technology?" *davidmeyercreations.com.*

Michell, John. *Secrets of the Stones: The Story of Astro-Archaeology.* Middlesex, UK: Penguin Books, Ltd., 1977.

———. *The New View over Atlantis.* San Francisco, California: Harper Collins, 1983.

———. *The View over Atlantis.* New York: Ballantine Books, 1972.

———. "Watkins's Revelation." Journal of Geomancy volume 3, number 1. October 1978.

Miller, Mike. "Coral Castle." Florida Backroads Travel. Updated November 25, 2020. *florida-backroads-travel.com.*

MissSquatcher. "The Infrasonic Effects of Bigfoots." Sasquatch Chronicles. November 18, 2015. *sasquatchchronicles.com/.*

"MoD Boscombe Down." Wikiwand. 2021. *wikiwand.com.*

Monroe, Robert A. *Journeys Out of the Body.* Garden City, New York: Anchor Books, 1973.

Mooney, Richard. *Colony Earth.* 1974.

Moore, William, and Charles Berlitz. *The Philadelphia Experiment: Project Invisibility.* Fawcett, 1995.

Muggenthaler, Elizabeth Von. "Low Frequency and Infrasonic Vocalizations from Tigers." Acoustics. December 6, 2000. *acoustics.org.*

"Nefertiti." World History. 2021. *worldhistory.org.*

"Nine Ladies Stone Circle." English Heritage. 2021. *english-heritage.org.uk.*

Noorbergen, Rene. *Secrets of the Lost Races: New Discoveries of Advanced Technology in Ancient Civilizations.* New York: The Bobbs-Merrill Company, Inc., 1977.

Noyes, Ralph. *A Secret Property.* London: Quartet Books, Ltd., 1985.

Nuwer, Rachel. "The Science of the Red Sea's Parting." *Smithsonian Magazine.* December 8, 2014. *smithsonianmag.com.*

Orme, Greg. "The Meridiani Face on Mars." *Journal of Space Exploration.* 2017.

———. *Why We Must Go to Mars: The King's Valley.* CreateSpace, 2011.

Ostrander, Sheila, and Lynn Schroeder. *Psychic Discoveries behind the Iron Curtain.* Englewood, New Jersey: Prentice-Hall, Inc., 1970.

"Overview and History." Rollright Stones. 2021. *rollrightstones.co.uk.*

Parry, Gareth. "Peace Women Fear Electronic Zapping at Base." *Guardian.* March 10, 1986.

Pearson, Mike Parker, Josh Pollard, Colin Richards, Kate Welham, Timothy Kinnaird, Dave Shaw, Ellen Simmons, Adam Standford, Richard Bevins, Rob Ixer, Clive Ruggles, Jim Rylatt, and Kevan Edinborough. "The Original Stonehenge? A Dismantled Stone Circle in the Preseli Hills of West Wales." *eprints.bournemouth.ac.uk.*

Petricevic, Ivan. "Unwritten Mystery: Did Ancient Civilizations Know the Secrets of Levitation?" Curiosmos. September 11, 2020. *curiosmos.com.*

"Philadelphia Experiment: A Conspiracy Widely Considered a Hoax." Unrevealed Files. April 14, 2020. *unrevealedfiles.com.*

Pilkington, Mark. "Ancient Electricity." *The Guardian.* April 21, 2004. *theguardian.com.*

———. "Flight of Fancy." *The Guardian.* June 30, 2005. *theguardian.com.*

Plackett, Benjamin. "Why Was Stonehenge Built?" LiveScience. January 9, 2021. *livescience.com.*

Prakash, Tara. "Egypt in the Old Kingdon (ca. 2649–2130 B.C.)." February 2019. *metmuseum.org.*

"Pre-Hispanic Town of Uxmal." UNESCO. 2021. *unesco.org.*

"Preseli Hills Pembrokeshire Wales." July 31, 2019. FelinFach. 2021. *felinfach.com.*

Puharich, Andrija. *The Sacred Mushroom: Key to the Door of Eternity.* CreateSpace Independent Publishing Platform, 2021.

The Quarterly Journal of Science. January 1874.

Radford, Ben. "Mystery of the Coral Castle Explained." Live Science. June 1, 2018. *livescience.com.*

———. "What Is the Taos Hum?" Live Science. February 19, 2014. *livescience.com.*

"RAF Greenham Common." 2020. *wikipedia.org.*

"Raytheon Technologies." LinkedIn company profile. *linkedin.com.*

"Red Sea." *New World Encyclopedia.* 2021. *newworldencyclopedia.org.*

Redfern, Nick. "A Strange and Controversial Saga from the Cold War." December 28, 2019. *mysteriousuniverse.org.*

———. "Commanding and Controlling the Weather by 2025?" Mysterious Universe. June 29, 2019. *mysteriousuniverse.org.*

———. Interview with Ray Boeche. January 22, 2007.

———. Interview with Linda Godfrey, April 6, 2003.

———. Interview with Mac Tonnies, July 7, 2009.

———. Interview with Mac Tonnies, March 14, 2004.

———. Interview with Mac Tonnies, September 9, 2006.

———. *Nessie!* Lisa Hagan Books, 2018. *amazon.com.*

———. *The Slenderman Mysteries: An Internet Urban Legend Comes to Life.* Red Wheel Weiser, 2017.

"Religious Levitation." *PSI Encyclopedia.* 2021. *psi-encyclopedia.spr.ac.uk.*

"Research on Stonehenge." English Heritage. 2021. *english-heritage.org.uk.*

Richards, Steve. *Levitation: The First Comprehensive Manual for Transcending Gravity through Meditational Techniques.* Wellingborough, UK: The Aquarian Press, 1980.

Richardson, Alan. *Spirits of the Stones: Visions of Sacred Britain.* London: Virgin, 2001.

Ripley. *Believe it or Not.* New York: Simon & Schuster, 1929.

"Robert M. Schoch." 2021. *sourcewatch.org.*

Robins, Don. *Circles of Silence.* London: Souvenir Press, 1985.

Robinson, Malcolm. *The Monsters of Loch Ness.* Lulu.com, 2016.

Ross, Anne, and Don Robins. *The Life and Death of a Druid Prince: The Story of Lindow Man, an Archaeological Sensation.* New York: Simon & Schuster, 1989.

Ross, David. "The Rollright Stones." Britain Express. 2021. *britainexpress.com.*

Roybal, Beth. "Sleep Paralysis." WebMD. October 17, 2020. *webmd.com.*

Satter, Raphael G. "UK. Experts Say Stonehenge Was a Place of Healing." *Seattle Times.* September 23, 2008. *seattletimes.com.*

Schoch, Robert M. "Research Highlights, the Great Sphinx." 2021. *robertschoch.com.*

Schwartz, Jean-Michel, and Lowell Bair. *The Mysteries of Easter Island: A Startling, Mind-Opening Investigation into Ancient Sources of Lost Knowledge.* New York: Avon Books, 1975.

Screeton, Paul. *Quest for the Hexham Heads.* Bideford, UK: Fortean Words, 2010.

Sellier, Charles E. *Mysteries of the Ancient World.* New York: Dell Books, 1995.

Serena, Katie. "These Supernatural Lines Supposedly Connect the Universe through Monuments and Landforms." All That's Interesting. March 14, 2018. *allthatsinteresting.com.*

Sethi, Ankit. "Coral Castle Mystery (Homestead)—Museum, Florida, Edward Leedskalnin." Mysterious Trip. Updated September 18, 2021. *mysterioustrip.com.*

Sitchin, Zecharia. "Baalbek: War Comes to the Landing Place." 2006. *www.bibliotecapleyades.net/.*

Smith, Warren. *The Secret Force of the Pyramids.* New York: Zebra Books, 1975.

Smythies, E.A., in David Hatcher Childress, *Technology of the Gods. books.google.com.*

"Sounds like Terror in the Air." Reuters. September 9, 2003. *smh.com.au.*

"Stanton Moor and Nine Ladies Stone Circle." 2021. *letsgopeakdistrict.co.uk.*

Stetka, Bret. "Sound Waves Can Heal Brain Disorders." *Scientific American.* 2014. *scientificamerican.com.*

"Stonehenge." English Heritage. *english-heritage.org.uk.*

"Stonehenge." History in Numbers. *historyinnumbers.com.*

"Stonehenge Bluestones Had Acoustic Properties, Study Shows." BBC News. March 3, 2014. *bbc.com.*

"The Stone Tape Theory." Haunted Walk. 2021. *hauntedwalk.com.*

"The Story: Story of England's Most Famous Prophetess." 2021. *mothershipton.co.uk.*

"The Strange Case of the Bristol Hum." BBC. January 19, 2016. *bbc.com.*

Summers, Montague. *The History of Witchcraft.* New York: University Books, Inc., 1965.

"Taos Hum, The." Astonishing Legends. October 15, 2020. *astonishinglegends.com.*

Tesla, Nikola. "The Wonder World to be Created by Electricity." 2021. *teslauniverse.com.*

Tingley, Brett. "The Boscombe Down Incident Remains One of Military Aviation's Most Intriguing Mysteries." The Drive. April 7, 2021. *thedrive.com.*

Tingley, Brett. "Scientists Achieve Force Field Levitation." Mysterious Universe. August 16, 2016. *mysteriousuniverse.org.*

Tomas, Andrew. *We Are Not the First: Riddles of Ancient Science*. London: Souvenir Press, Ltd., 1973.

Tonnies, Mac. *After the Martian Apocalypse*. Pocket Books, 2004.

Toth, Max, and Greg Nielsen. *Pyramid Power: The Secret of the Ancients Revealed . . . The World's Greatest Mystery.* 1976.

Trevelyan, Marie. *Folk-Lore and Folk-Stories of Wales*. Whitefish, Montana: Kessinger Publishing, LLC, 2007.

Trick, Julian. "Science in Search of the Low Rumble." Free Public. October 9, 2002. *freerepublic.com.*

Twain, Mark. *The Innocents Abroad.* CreateSpace Independent Publishing Platform, 2018.

"Twin Peaks in 3-D, as Viewed by the Mars Pathfinder IMP Camera, The." NASA. November 4, 1997. *nasa.gov.*

University of Bristol. "Now You Can Levitate Liquids and Insects at Home." ScienceDaily. August 15, 2017. *sciencedaily.com.*

"USNS Eltanin (T-AK-270)." Military Wiki. *military-historyfandom.com.*

Valentine, Tom. *The Great Pyramid: Man's Monument to Man*. New York: Pinnacle Books, Inc., 1975.

Volpe, Sam. "Highgate 'Vamipre Hunter' Dies Half a Century after Supernatural Panic Gripped Community." Ham&High. April 24, 2019, updated October 14, 2020. *hamhigh.co.uk.*

Wagner, Stephen. "Levitation: A Forgotten Art Used to Build Monuments." LiveAbout. Updated February 17, 2019. *liveabout.com.*

Wedeck, Harry E. *A Treasury of Witchcraft.* Bonanza Books, 1961.

"What Is Causing the Taos Hum?" June 8, 2019. *curioushistorian.com.*

"What Is the Importance of the Parting of the Red Sea?" Got Questions. *gotquestions.org.*

"What Is the Meaning and Importance of the Exodus from Egypt?" Got Questions. *gotquestions.org.*

"White Powder of Gold (ORME)." Token Rock. 2021. *tokenrock.com.*

"Who's Ed?" Coral Castle. 2018. *coralcastle.com.*

“Why Do Whales Make Sounds?” National Oceanic and Atmospheric Administration. February 26, 2021. *noaa.gov.*

“William Stainton Moses.” Theosophy Wiki. *theosophy.wiki.*

Wilson, Tracy V. “How Acoustic Levitation Works.” How Stuff Works. 2021.

Winston, Steve. “Homestead’s Coral Castle: Monument to a Lost Love.” *Philadelphia Inquirer.* March 20, 2012. *inquirer.com.*

“The Wonder World to Be Created by Electricity.” September 9, 1915. *pbs.org.*

Zink, Dr. David. *The Stones of Atlantis.* Englewood Cliffs, New Jersey: Prentice-Hall, Inc., 1978.

ACKNOWLEDGMENTS

I would like to thank everyone at Red Wheel Weiser and particularly Greg Brandenburgh, Michael Pye, Laurie Kelly, Kathryn Sky-Peck, Eryn Carter, and Jodi Brandon; and my literary agent and good friend, Lisa Hagan, for her tireless work; Linda Godfrey for a great interview; and Kimberly Rackley, without whose expertise in the field of remote-viewing significant portions of this book could never have been written.

ABOUT THE AUTHOR

Nick Redfern is the author of more than sixty books, including *The Martians, Keep Out!, For Nobody's Eyes Only, Immortality of the Gods, Top Secret Alien Abduction Files, Men in Black, Bloodline of the Gods, Contactees, The Pyramids and the Pentagon,* and *Weapons of the Gods.* He has been on many television shows, including Travel Channel's *In Search of Monsters,* History Channel's *UnXplained* and *Monster Quest,* SyFy Channel's *Proof Positive,* and the National Geographic Channel's *Paranatural.* Nick is a regular guest on *Coast to Coast AM.*

Nick's blog: *World of Whatever, nickredfernfortean.blogspot.com*
Nick's Twitter account: @nickredfernufo
Nick's Facebook page: *www.facebook.com/nick.redfern.73*